Advances in Computer Vision and Pattern Recognition

For further volumes:
http://www.springer.com/series/4205

Bogdan Ionescu · Jenny Benois-Pineau
Tomas Piatrik · Georges Quénot
Editors

Fusion in Computer Vision

Understanding Complex Visual Content

Editors
Bogdan Ionescu
University Politehnica of Bucharest
Bucharest
Romania

Jenny Benois-Pineau
University of Bordeaux
Talence
France

Tomas Piatrik
Queen Mary University of London
London
UK

Georges Quénot
Laboratory of Informatics of Grenoble
Grenoble
France

Series editors
Sameer Singh
Rail Vision Europe Ltd.
Castle Donington
Leicestershire
UK

Sing Bing Kang
Interactive Visual Media Group
Microsoft Research
Redmond, WA
USA

ISSN 2191-6586 ISSN 2191-6594 (electronic)
Advances in Computer Vision and Pattern Recognition
ISBN 978-3-319-05695-1 ISBN 978-3-319-05696-8 (eBook)
DOI 10.1007/978-3-319-05696-8
Springer Cham Heidelberg New York Dordrecht London

Library of Congress Control Number: 2014933571

© Springer International Publishing Switzerland 2014
This work is subject to copyright. All rights are reserved by the Publisher, whether the whole or part of the material is concerned, specifically the rights of translation, reprinting, reuse of illustrations, recitation, broadcasting, reproduction on microfilms or in any other physical way, and transmission or information storage and retrieval, electronic adaptation, computer software, or by similar or dissimilar methodology now known or hereafter developed. Exempted from this legal reservation are brief excerpts in connection with reviews or scholarly analysis or material supplied specifically for the purpose of being entered and executed on a computer system, for exclusive use by the purchaser of the work. Duplication of this publication or parts thereof is permitted only under the provisions of the Copyright Law of the Publisher's location, in its current version, and permission for use must always be obtained from Springer. Permissions for use may be obtained through RightsLink at the Copyright Clearance Center. Violations are liable to prosecution under the respective Copyright Law.
The use of general descriptive names, registered names, trademarks, service marks, etc. in this publication does not imply, even in the absence of a specific statement, that such names are exempt from the relevant protective laws and regulations and therefore free for general use.
While the advice and information in this book are believed to be true and accurate at the date of publication, neither the authors nor the editors nor the publisher can accept any legal responsibility for any errors or omissions that may be made. The publisher makes no warranty, express or implied, with respect to the material contained herein.

Printed on acid-free paper

Springer is part of Springer Science+Business Media (www.springer.com)

It was six men of Indostan
To learning much inclined,
Who went to see the Elephant
(Though all of them were blind),
That each by observation
Might satisfy his mind.

—John Godfrey Saxe

Preface

Understanding of Complex Visual Content is essential for a wide range of important applications including automatic multimedia content indexing and retrieval, medicine, robotics, or surveillance. This is a difficult problem due to what we call the "semantic gap" or the distance between the raw representation of image or video contents (bit streams) and the concepts and relations between them that are meaningful and useful for human beings.

Many approaches rely on the joint use of content representation and supervised machine learning techniques though the recent approaches like deep learning now attempt to do both at once. Many alternative and complementary content representation and machine learning techniques now exist. While some single or elementary representation/classification combinations perform relatively well, none of them is currently able to capture and fully exploit the raw media stream for understanding its contents. Research is still trying to explore the best single combinations and their improvements in several complementary directions and the fusion of such elementary combinations is a way of further improving the overall system performance.

This book focuses on the fusion problem in a variety of domains and applications. It follows the workshop on Information Fusion in Computer Vision for Concept Recognition held jointly with the 12th European Conference on Computer Vision (ECCV2012). It contains extended versions of works presented in this workshop together with other works carried out by leading researchers in the domain. The different chapters cover many aspects of the problem and describe successful approaches evaluated in the context of international benchmarks that model realistic use cases at significant scales.

Visual and multi-modal scene understanding by humans is a result of high level interpretation of quantities of information we gather by different physiological channels. We are sensitive to colors, contrasts, motion, visual "roughness," "granularity," loudness of sounds and their nature, etc. We are fusing these different sources to recognize and understand the content. In computer vision and multimedia nowadays, we are imitating this fusion process at different levels. We are speaking of "early," "late," and "intermediate" fusion for scene understanding.

Under "early fusion," we usually understand building rich feature spaces for content description as well as the transformation of these spaces to get the highest efficiency in the content recognition task.

The "late fusion" term denominates the fusion of results of primary decisions, often using information from a single description subspace. The various combination operators, including cascade classification approaches, are applied for aggregating primary decisions in the overall recognition task.

In the "intermediate fusion" approaches we combine results obtained on description subspaces, often coming from different modalities.

Today, the research community in computer vision and multimedia can benchmark their methods on large datasets in the scope of evaluation campaigns, such as ImageCLEF, TRECVid, Pascal VOC, MediaEval, etc. These competitions show the interest and ever-growing performances of late fusion schemes.

The book is organized as follows. In Chaps. 1–3 we are interested in the late fusion approaches for concept recognition in images and videos. A specific accent is made on the study, in Chap. 2, of a very popular model of visual content, namely Bag-of-(Visual)-Words and various fusion aspects which are analyzed in this framework.

Chapter 4 presents an ever-growing trend in the interpretation of visual content by incorporating models of Human Visual System with content understanding methods. Here we are also speaking about fusion. To delimit the areas of potential attention in the so-called bottom-up image-driven manner, multiple cues have to be fused: motion, contrast, and geometry of scenes. The approach is also incorporated in the classical Bag-of-(Visual)-Words model.

In Chap. 5 fusion schemes are developed for a more focused task, such as example-based event recognition in video. Multi-modal features of different semantic levels, such as Bag-of-(Visual)-Words, motion features, audio features, but also results of semantic concepts detections are fused together to recognize events of interest. The interesting conclusion on a good performance of simple fusion operators, such as linear combination of intermediate classification results, is given.

Analyzing a very rich state-of-the art research in computer vision in the matter of scene understanding, one can roughly say that the most efficient approaches follow a threefold scheme: content description, classification, and fusion. All of them are important, nevertheless, the classification approach is the core. In Chap. 6, rotation-based ensemble classifiers for high-dimensional data are proposed, which encourage both individual accuracy and diversity within the ensemble simultaneously.

Chapters 7–9 are more application-focused and present the search of optimal strategies of fusion in such applications as video surveillance, violent content detection in movies, and biomedical information retrieval.

Information fusion is a model of human interpretation of complex visual content. Nevertheless, we are very far from saying today that the mechanisms of content understanding by humans are fully explored. We are only at the beginning in modeling the process. This is why we need to study on a large scale how humans interpret the content. The last Chap. 10 of the book is devoted to this key question.

We dedicate this book to researchers and students working in the domain of information fusion for complex visual contents understanding or working in other related domains where mentioned techniques can be applied.

We are deeply grateful to all those who have helped us to successfully organize the workshop and to produce this book. Special thanks to the reviewers who have contributed to the high quality of the work presented here. We are indebted to all authors for their contribution and hope that the readers of the book will appreciate their hard work and enjoy the reading. Finally, we thank our editor, Springer, who gave us the opportunity to bring this project to life.

Bogdan Ionescu
Jenny Benois-Pineau
Tomas Piatrik
Georges Quénot

Acknowledgments

The book was produced with the kind support of the French National Research Network in Information, Signal, Image and Vision (GDR-ISIS).

Contents

1 **A Selective Weighted Late Fusion for Visual Concept Recognition**................................ 1
Ningning Liu, Emmanuel Dellandréa, Bruno Tellez and Liming Chen

2 **Bag-of-Words Image Representation: Key Ideas and Further Insight**................................ 29
Marc T. Law, Nicolas Thome and Matthieu Cord

3 **Hierarchical Late Fusion for Concept Detection in Videos**...... 53
Sabin Tiberius Strat, Alexandre Benoit, Patrick Lambert, Hervé Bredin and Georges Quénot

4 **Fusion of Multiple Visual Cues for Object Recognition in Videos**... 79
Iván González-Díaz, Jenny Benois-Pineau, Vincent Buso and Hugo Boujut

5 **Evaluating Multimedia Features and Fusion for Example-Based Event Detection**...................... 109
Gregory K. Myers, Cees G. M. Snoek, Ramakant Nevatia, Ramesh Nallapati, Julien van Hout, Stephanie Pancoast, Chen Sun, Amirhossein Habibian, Dennis C. Koelma, Koen E. A. van de Sande and Arnold W. M. Smeulders

6 **Rotation-Based Ensemble Classifiers for High-Dimensional Data**.. 135
Junshi Xia, Jocelyn Chanussot, Peijun Du and Xiyan He

7 **Multimodal Fusion in Surveillance Applications**............. 161
Virginia Fernandez Arguedas, Qianni Zhang and Ebroul Izquierdo

8 **Multimodal Violence Detection in Hollywood Movies: State-of-the-Art and Benchmarking**............................. 185
Claire-Hélène Demarty, Cédric Penet, Bogdan Ionescu,
Guillaume Gravier and Mohammad Soleymani

9 **Fusion Techniques in Biomedical Information Retrieval**........ 209
Alba García Seco de Herrera and Henning Müller

10 **Using Crowdsourcing to Capture Complexity in Human Interpretations of Multimedia Content**........................ 229
Martha Larson, Mark Melenhorst, María Menéndez and Peng Xu

Index... 271

Chapter 1
A Selective Weighted Late Fusion for Visual Concept Recognition

Ningning Liu, Emmanuel Dellandréa, Bruno Tellez and Liming Chen

Abstract We propose a novel multimodal approach to automatically predict the visual concepts of images through an effective fusion of visual and textual features. It relies on a Selective Weighted Late Fusion (SWLF) scheme which, in optimizing an overall Mean interpolated Average Precision (MiAP), learns to automatically select and weight the best features for each visual concept to be recognized. Experiments were conducted on the MIR Flickr image collection within the ImageCLEF Photo Annotation challenge. The results have brought to the fore the effectiveness of SWLF as it achieved a MiAP of 43.69 % in 2011 which ranked second out of the 79 submitted runs, and a MiAP of 43.67 % that ranked first out of the 80 submitted runs in 2012.

1.1 Introduction

Over the last few years, we have witnessed an explosion in the quantity of multimedia data available both for professional and personal uses. This has been amplified by technological advances realized in various domains that not only address professionals but also consumers such as digital content creation, diffusion, and storage. Consequently, more and more multimedia documents are stored on personal

---

N. Liu (✉) · E. Dellandréa · B. Tellez · L. Chen
Université de Lyon, CNRS, Ecole Centrale de Lyon, LIRIS,
UMR5205, Lyon 69134, France
e-mail: ningning.liu@ec-lyon.fr
http://liris.cnrs.fr/

E. Dellandréa
e-mail: emmanuel.dellandrea@ec-lyon.fr

B. Tellez
e-mail: bruno.tellez@univ-lyon1.fr

L. Chen
e-mail: liming.chen@ec-lyon.fr

computers but also on the Internet. For example, one of the famous online photo sharing websites, Flickr[1] deals with an average of 1.42 million photos uploaded each day, hosting more than 7 billion photos. Meanwhile, one of the famous social networking websites, Facebook[2] serves a peak of 1.2 million photos per second, hosting more than 170 billion images. Basically, by adding text descriptions or keywords to the multimedia data, it becomes possible to retrieve those documents by using classical text-based retrieval techniques. However, nowadays, those text-based systems are no more conceivable due to the following drawbacks:

- As the amount of data is huge, it is not possible to efficiently annotate the data manually.
- As the manual annotation is subjective, it is not guaranteed to obtain coherent descriptions from persons with different backgrounds. Meanwhile, users tend to use only the first few words that come to their mind, thus, the annotation is incomplete.
- As the text descriptions are only available with a limited number of languages, this would make the choice of a proper language more critical for annotation and search, or introduce more noise by using unstable automatic translation techniques.

To some extent, these problems might be overcome by a more careful manual indexing. This however, would be associated with more effort and is infeasible in most situations. In such context, to escape the limits of manual tagging and annotating, content-based image retrieval (CBIR) systems spring up and support image search based on low-level visual features, such as colors, textures, or shapes [1, 2]. However, human perception and understanding of images are subjective and rather on the concept level [3–7]. Therefore, the research topic of visual concept recognition has attracted increasing attention from researchers including psychology, computer science, linguistics, neuroscience, and related disciplines, which could greatly be beneficial to the mentioned applications.

Machine-based recognition of visual concepts aims at automatically recognizing high-level visual semantic concepts (HLSC), including scenes (such as indoor, outdoor, landscape, etc.), objects (car, animal, person, etc.), events (travel, work, etc.), or even emotions (melancholic, happy, etc.). It proves to be extremely challenging because of high inter-class similarities and large intra-class variations especially due to appearance deformations, occlusions, background clutter, changes in viewpoint, pose, scale and illumination. The past decade has witnessed tremendous efforts from the research communities as testified by the multiple challenges in the field, such as Pascal VOC [8], TRECVID [9] and ImageCLEF [10]. Most approaches to visual concept recognition (VCR) have so far focused on appropriate visual content description, and have featured a dominant bag-of-visual-words (BoVW) representation along with local SIFT descriptors. Meanwhile, more and more works in the literature have discovered the wealth of semantic meanings conveyed by the abundant textual captions associated with images [11]. Therefore, multimodal approaches are

---

[1] http://www.flickr.com
[2] http://www.facebook.com

proposed for VCR by making joint use of user textual tags and visual descriptions to bridge the gap between HLSC and low-level visual features. Our work is in that line and targets an effective feature fusion scheme for VCR.

The rest of this chapter is organized as follows. We present in Sect. 1.2 a review of the state of the art related to multimodal visual concept recognition. The proposed fusion scheme, SWLF, is presented in Sect. 1.3. Section 1.4 presents the features including visual and textual representations. The experiments we have conducted to evaluate SWLF are described in Sect. 1.5. Finally, we give the conclusion in Sect. 1.6.

1.2 Related Work

As far as multimodal approaches are concerned, it requires a fusion strategy to combine information from multiple sources, e.g., visual stream and sound stream for video analysis [12], textual, and visual content for multimedia information retrieval [13], etc. Taking the multimodal approaches for VCR as examples, there are generally two modalities to handle, namely the textual and the visual modality. A fusion step is required to obtain the final results based on the analysis of different modalities, in which different types of features can be computed to form several information streams. These streams need to be fused in order to elaborate a final decision. Fusion methods can be classified into the following three strategies:

- Early fusion: All the features from the different modalities are simply concatenated into one single feature vector that is provided as input of the classifier to obtain the final classification result, as in [14].
- Late fusion: The features from each individual channel are first used to feed a classifier to get its classification score. The scores from all the channels are then combined to obtain the final classification score according to a certain criterion, such as mean, max, min, and weighted sum, as in [15]. A comparison of early and late fusion strategies is given in Fig. 1.1.
- Intermediate fusion: This concerns fusion strategies that are neither early nor late, as in [18]. For instance, Noble's approach [16], Pinquier's HMM-based approach [17], and more generally Multiple Kernel Learning (MKL) algorithms that consist in combining kernels from each type of feature, falls into this category.

Each kind of fusion strategy has its own advantages and drawbacks. The early fusions are straightforward and simple by just concatenating the features extracted from various information sources into a single vector representation, and their drawbacks are their large dimensionality and the difficulty they encounter in combining features of different natures into a common homogeneous representation. The intermediate fusions, mostly kernel methods, present good performances on different visual classification tasks [8, 19–21], but their weakness lies in their high computation cost with a number of parameters to learn. By contrast, the late fusion strategies, which consist in integrating the scores of the classifiers outputs, not only provide a trade-off between preservation of information and computational efficiency but also

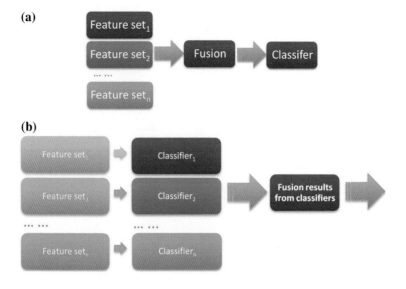

Fig. 1.1 A comparison between early (**a**) and late (**b**) fusion strategies

prove to perform speedily compared to early fusions [22, 23]. A comprehensive and comparative study of various combination rules, such as sum, product, max, min, median, and majority voting, by Kittler et al. [24], suggests that the sum rule is much less sensitive to the error of individual classifiers when estimating posterior class probability.

The proposed fusion scheme described in the next section, namely Selective Weighted Late Fusion (SWLF), falls into the category of late fusion strategies. Specifically, when different features such as visual and textual ones, can be used for VCR, SWLF learns to automatically select and weight the best features to be fused for each visual concept to be recognized.

1.3 Selective Weighted Late Fusion

The fusion scheme that we propose is a selective weighted late fusion (SWLF), which shares the same idea as the adaptive score level fusion scheme proposed by Soltana et al. [25]. Meanwhile, recently, Pinquier et al. [17] proposed a optimal fusion strategies in the Hierarchical Hidden Markov Model (HMM) framework for activities recognition, which is also based on weighted trusted experts. While a late fusion at score level is reputed as a simple and effective way to fuse features of different nature for machine-learning problems, the proposed SWLF builds on two simple insights. First, the score delivered by a feature type should be weighted by its intrinsic quality for the classification problem at hand. Second, in a multi-label

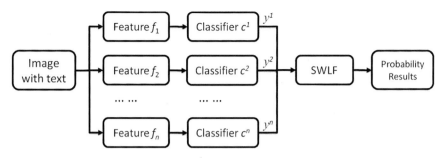

Fig. 1.2 The framework of the SWLF scheme. For each image and each concept, the associated tags are analyzed to extract the textual features for textual classifiers. Meanwhile, visual features are extracted to feed visual classifiers. Experts (classifiers) are then combined to predict the presence of a given concept in the input image

scenario where several visual concepts may be assigned to an image, different visual concepts may require different features that best recognize them. For instance, the "sky" concept may greatly require global color descriptors, while the best feature to recognize a concept like street could be a segment-based feature for capturing straight lines of buildings. The whole SWLF framework is illustrated in Fig. 1.2.

1.3.1 The Principle of SWLF

The proposed SWLF scheme has a learning phase which requires a training dataset for the selection of the best experts and their corresponding weights for each visual concept. Specifically, given a training dataset, we divide it into two disjoint parts composed of a training set and a validation set. For each visual concept, a binary classifier (one versus all) is trained, which is also called *expert* in the subsequent, for each type of features using the data in the training set. Thus, for each concept, we generate as many experts as the number of different types of features. The quality of each expert can then be evaluated through a quality metric using the data in the validation set. In this work, the quality metric is chosen to be the interpolated Average Precision (iAP), which is computed so that the recall measurements range from 0.0 to 1.0 interpolated with steps of 0.1. The higher iAP is for a given expert, the more weight should be given to the score delivered by that expert for the late fusion. Concretely, given a visual concept k, the quality metrics, i.e. iAP, produced by all the experts are first normalized into w_k^i. To perform a late fusion of all these experts at score level, the *sum of weighted scores* is then computed as in (1.1):

$$\text{score} : z_k = \sum_{i=1}^{N}(w_k^i * y_k^i), \qquad (1.1)$$

where y_k^i represents the score of the ith expert for the concept k, and w_k^i stands for the normalized iAP performance of the feature f_i on the validation dataset. In the subsequent, late fusion through (1.1) is called *weighted score* rule.

For the purpose of comparison, we also consider three other score level fusion schemes, namely "min," "max," or "sum" rules that are recalled respectively in Eqs. (1.2)–(1.4):

$$\min : z_k = \min(y_k^1, y_k^2, \ldots, y_k^N); \quad (1.2)$$

$$\max : z_k = \max(y_k^1, y_k^2, \ldots, y_k^N); \quad (1.3)$$

$$\text{mean} : z_k = \frac{1}{N} \sum_{i=1}^{N} y_k^i; \quad (1.4)$$

Actually, these three fusion rules can have very simple interpretation. The *min* fusion rule is the consensus voting. A visual concept is recognized only if all the experts recognize it. The *max* rule can be called alternative voting. A visual concept is recognized as long as one expert has recognized it. Finally, the *mean* rule can be assimilated as the majority voting where a concept is recognized if the majority of the experts recognize it.

In practice, one discovers that the late fusion of all the experts leads to a decrease in the global classification accuracy, i.e., the mean iAP over the whole set of visual concepts. The reason could be that some of the features so far proposed can be noisy and irrelevant to a certain number of visual concepts, thus disturbing the learning process and lowering the generalization skill of the learnt expert on the unseen data. For this purpose, we further implement the SWLF scheme inspired by a wrapper feature selection method, namely the SFS method (Sequential Forward Selection) [26], which first initializes an empty set, and at each step the feature that gives the highest correct classification rate along with the features already included is added to the set of selected experts to be fused. More specifically, for each visual concept, all the experts are sorted in a decreasing order according to their iAP. At a given iteration N, the only first N experts are used for late fusion and their performance is evaluated over the data of the validation set. N is increased until the overall classification accuracy measured in terms of MiAP starts to decrease. The learning procedure of the SWLF algorithm can be defined as follows:

Selective weighted late fusion (SWLF) algorithm for training

Input: Training dataset T (of size N_T) and validation dataset V (of size N_V).
Output: Set of N experts for the K concepts $\{C_k^n\}$ and the corresponding set of weights $\{\omega_k^n\}$ with $n \in [1, N]$ and $k \in [1, K]$.
Initialization: $N = 1$, $MiAP_{\max} = 0$.

- Extract M types of features from T and V
- For each concept $k = 1$ to K
  - For each type of feature $i = 1$ to M
    1. Train the expert C_k^i using T
    2. Compute ω_k^i as the iAP of C_k^i using V
  - Sort the ω_k^i in descending order and denote the order as $j^1, j^2, \ldots, j^M$ to form $W_k = \{\omega_k^{j^1}, \omega_k^{j^2}, \ldots, \omega_k^{j^M}\}$ and the corresponding set of experts $E_k = \{C_k^{j^1}, C_k^{j^2}, \ldots, C_k^{j^M}\}$
- For the number of experts $n = 2$ to M
  - For each concept $k = 1$ to K
    1. Select the first n experts from E_k: $E_k^n = \{C_k^1, C_k^2, \ldots, C_k^n\}$
    2. Select the first n weights from W_k: $W_k^n = \{\omega_k^1, \omega_k^2, \ldots, \omega_k^n\}$
    3. For $j = 1$ to n: Normalize $\omega_k^{j'} = \omega_k^j / \sum_{i=1}^n \omega_k^i$
    4. Combine the first n experts into a fused expert, using the *weighted score* rule through
       (1.1): $z_k = \sum_{j=1}^n \omega_k^{j'} \cdot y_k^j$ where y_k^j is the output of C_k^j
    5. Compute $MiAP_k^n$ of the fused expert on the validation set V
  - Compute $MiAP = 1/K \cdot \sum_{k=1}^K MiAP_k^n$
  - If $MiAP > MiAP_{\max}$
    - Then $MiAP_{\max} = MiAP, N = n$
    - Else break

As a late fusion strategy, the computational complexity of SWLF can be computed in terms of the number of visual concepts, K and the number of types of features, M. This complexity is $O(K \times M^2)$. Note that the optimized fusion strategy achieved through SWLF only needs to be trained once on the training and validation datasets.

SWLF combines an ensemble of experts for a better prediction of class labels, i.e., visual concepts in this work. From this perspective, SWLF can also be viewed as a method of ensemble learning [27] which aims to use multiple models to achieve better predictive performance than could be obtained from any of the constituent models. Nevertheless, SWLF differs from popular bagging methods [28], such as random forest, which involve having each expert in the ensemble trained using a randomly drawn subset of a training set and vote with equal weight. In the case of SWLF, the training dataset is divided into a training set and a validation set which are used to train experts and SWLF to select the best ones for fusing using different weights.

1.3.2 The Variants of SWLF

As the number of experts N is the same for each concept in the above algorithm, this version of SWLF is called *SWLF_FN* (fixed N). However, several variants can be built upon SWLF. Indeed, instead of fixing the same number of experts N for all concepts, it is possible to select the number of experts on a per-concept basis. Thus the number of experts can be different for each concept. Therefore, we have implemented this variant denoted *SWLF_VN* (variable N) in the following. In this case, for each concept, features are incrementally selected until the iAP for this concept begins to decrease. Another variant concerns the way the experts are selected at each iteration. Instead of adding the nth best expert at iteration n to the set of previously selected $n - 1$ experts, one can also select the expert that yields the best combination of n experts, in terms of $MiAP$, once added to the set of $n - 1$ experts already selected at the previous iteration. This variant is denoted *SWLF_SFS* in the following as the selection scheme is inspired from the feature selection method "Sequential Feature Selection" [26] generally used for early fusion of features.

1.4 The Features Used for VCR

More and more, images are shared on the Internet together with textual resources such as EXIF data, legends, or tags. This is for instance the case for Flickr website,[3] which is the data source of the ImageCLEF photo annotation challenge. In fact, these textual descriptions have proven to be a rich source of semantic information for the purpose of image classification and retrieval [29–31]. Therefore, in order to describe images for further classification, we propose to use not only visual features extracted from the image, but also textual features extracted from the textual resources associated with images. These features are briefly presented in the following sections.

1.4.1 Visual Features

As the concepts to be detected in images can be characterized by different visual properties, we propose to extract a rich set of features including low-level features based on color, texture, shape, being local or global, as well as mid-level features related to aesthetic and affective image properties.

In order to capture the global ambiance and layout of an image, we further compute a set of global features, including descriptions of color information in the HSV (Hue, Saturation, Value) color space in terms of means, color histograms and color moments, textures in terms of LBP (Local Binary Patterns) [32], Color LBP [33], co-occurrence and auto-correlation, as well as shape information in terms of histograms of line orientations quantized into 12 different orientations and computed by the Hough transform [34]. To describe image local content, we follow the dominant BoVW approach that views an image as an unordered distribution of local image features extracted from salient image points, called "interest points" [35, 36] or more simply from points extracted on a dense grid [37, 38]. In this work, we make use of several popular local descriptors, including C-SIFT, RGB-SIFT, HSV-SIFT [39], and DAISY [40], extracted from a dense grid [41]. An image is then modeled as a BoVW using a dictionary of 4,000 visual words and hard assignment. The codebook

---

[3] http://www.flickr.com/

size, 4,000 in this work, results from a trade-off between computational efficiency and the performance over a training dataset. The visual words represent the centers of the clusters obtained from the k-means algorithm. In addition to these local and global low-level features, we also collect and implement a set of mid-level features [42–44] that are mostly inspired from studies in human visual perception, psychology [45], cognitive science, art [46], etc., thus in close relationships with the 9 sentiment concepts newly introduced in the image annotation task at ImageCLEF 2011. These mid-level features include emotion-related visual features, aesthetic and face-related features.

In total, we extract 24 visual feature sets of various dimensions ranging from 1 for color harmony to 4,000 for each of the SIFT variants. Table 1.1 summarizes all the visual features used in the following experiments.

1.4.2 Textual Features

The last few years have seen an impressive growth of sharing websites particularly dedicated to videos and images. The famous Flickr website[4] for example, from which is extracted the MIR FLICKR image collection, allows users to upload and share their images and to provide a textual description under the form of tags or legends. These textual descriptions are a rich source of semantic information on visual data that is interesting to consider for the purpose of VCR or multimedia information retrieval. In [7], we have proposed a novel textual descriptor, namely the Histogram of Textual Concepts (HTC) that is inspired from the componential space model and which describes the meaning of a word by its atoms, components, attributes, behavior, related ideas, etc. Specifically, the HTC of a text document is defined as a histogram of textual concepts toward a vocabulary or dictionary, and each bin of this histogram represents a concept of the dictionary, whereas its value is the accumulation of the contribution of each word within the text document toward the underlying concept according to a predefined semantic similarity measure. Given a dictionary D and a semantic similarity measurement S, HTC can be simply extracted from the tags of an image through a three-step process as illustrated in Fig. 1.3. Note that the tags such as "peacock," "bird," "feathers," and "animal" all contribute to the bin values associated with the "animal" and "bird" concepts according to a semantic similarity measurement whereas the tags such as "beautiful," "pretty," and "interestingness" all help peak the bin value associated with the concept "cute". This is in clear contrast to the BoW approaches where the relatedness of textual concepts is simply ignored as word terms are statistically counted.

The advantages of HTC are multiple. First, for a sparse text document as image tags, HTC offers a smooth description of the semantic relatedness of user tags over a set of textual concepts defined within the dictionary. More importantly, in the case of polysemy, HTC helps disambiguate textual concepts according to the context.

---

[4] http://www.flickr.com/

Table 1.1 Summary of the visual features

Category	Short name	#	Short description
Color	grey_hist	128	128-bin histogram computed from the gray level image
	color_hsv	132	Concatenation of the 64-bin histograms computed from each HSV channel
	color_moment	144	3 central moments (mean, standard deviation and skewness) on HSV channels using a pyramidal image representation
	color_mSB	2	Mean saturation and brightness in HSV color space
Texture	texture_lbp	256	Standard LBP features [32]
	texture_tamura	3	Tamura features [47] including coarseness, contrast, directionality
	texture_cooccu	16	Distribution of co-occurring values in the image at a given offset [48]
	texture_autocorr	132	Autocorrelation image coefficients [49]
	hsvLbp invLbp rgbLbp oppoLbp	1311 (each)	Four multi-scale color LBP operators based on different color spaces [33]
Shape	shape_histLine	12	Histogram of 12 different orientations by using Hough transform [34]
Local descriptors	c-sift rgb-sift hsv-sift oppo-sift	4000 (each)	Four SIFT descriptors based on different color spaces and computed on a dense grid [39, 50]
	daisy	4000	DAISY descriptor computed on a dense grid [41]
Mid-level	mlevel_PAD	3	Emotional coordinates based on HSV color space according to [45]
	mlevel_harmony	1	Color harmony of images based on Itten's color theory [46]
	mlevel_dynamism	1	Ratio between the numbers of oblique lines in images with respect to the total number of lines [51, 52]
	mlevel_aesthetic YKe	5	Ke et al. [53] aesthetic criteria including: spatial distribution of edges, hue count, blur, contrast, and brightness
	mlevel_aesthetic Datta	44	Most of the features (44 of 56) except those that are related to IRM (integrated region matching) technique [54]
	mlevel_facect	5	Number of faces in the image detected by using five different pose configurations of the face detector from [55]

1 A Selective Weighted Late Fusion for Visual Concept Recognition

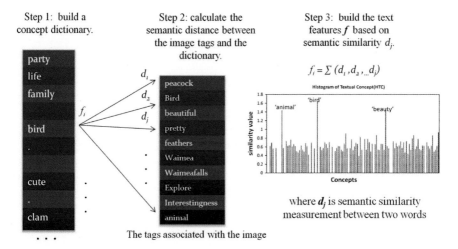

Fig. 1.3 The three-steps process of our HTC algorithm

Our experiments have further indicated that HTC features perform better than the dominant BoW approach (such as TF-IDF, LDA). The reason may lie in the fact that BoW-kind approaches assume that word terms are basically statistically independent, thereby mismatching text documents close in content but with different term vocabulary.

As the computation of HTC relies on the definition of a dictionary and the semantic distance measurement over textual concepts, we can build variants of HTC features by using different dictionaries and distance measurements.

In particular, in this work, we have implemented and compared two measurements of semantic similarities between two textual concepts, namely the *path* and the *wup* distances [56] that are based on the WordNet ontology [57]. Given two synsets w_1 and w_2, the *path* and the *wup* distances are defined as:

$$d_{\text{path}}(w_1, w_2) = \frac{1}{1 + \text{spl}(w_1, w_2)} \quad (1.5)$$

$$d_{\text{wup}}(w_1, w_2) = \frac{2 \times \text{depth}(\text{lcs}(w_1, w_2))}{\text{depth}(w_1) + \text{depth}(w_1)} \quad (1.6)$$

where lcs(w_1, w_2) denotes the least common subsumer (most specific ancestor node) of the two synsets w_1 and w_2 in the WordNet taxonomy, depth(w) is the length of the path from w to the taxonomy root, and spl(w_1, w_2) returns the distance of the shortest path linking the two synsets (if one exists). Note that the *path* and the *wup* measurements have opposite polarity. When the two synsets w_1 and w_2 are identical, *path* returns 1 while *wup* returns 0. Therefore, when using *wup* for accumulating the

Table 1.2 Different variants of the textual features based on HTC

Feature name	Dictionary	Similarity measure	Accumulating method
txtf_99ps	D_99	path	$f_i = \sum_t S(t,i)$
txtf_99pm	D_99	path	$f_i = \max_t S(t,i)$
txtf_99ws	D_99	wup	$f_i = \sum_t S(t,i)$
txtf_99wm	D_99	wup	$f_i = \max_t S(t,i)$
txtf_1034ps	D_Anew	path	$f_i = \sum_t S(t,i)$
txtf_1034pm	D_Anew	path	$f_i = \max_t S(t,i)$
txtf_1034ws	D_Anew	wup	$f_i = \sum_t S(t,i)$
txtf_1034wm	D_Anew	wup	$f_i = \max_t S(t,i)$
txtf_1034pva	D_Anew	path	$f_i = \max_t S(t,i)$
txtf_1034wva	D_Anew	wup	$f_i = \max_t S(t,i)$

Their names are related to the way they are computed. For instance, *txtf_99ps* refers to the HTC variant using the dictionary *D_99* made of ImageCLEF 2011 concept names along with the *path* distance as semantic similarity measurement, and the *sum* accumulating operator. *txtf_1034pvad* refers to the valence, arousal, and dominance coordinates, while the underlying HTC variant is computed using ANEW vocabulary *D_Anew* and the *path* distance. The *path* and *wup* distance are based on the WordNet ontology [57]

semantic similarities in the computation of HTC, its polarity is first changed to a positive one in our work.

Table 1.2 summarizes all the text features used in following experiments.

1.5 Experiments and Results

In order to make a comparison of our method with those among the most recent ones in the visual concept recognition domain, we carried out extensive experiments on the MIR FLICKR image collection [58, 59] that was used within the ImageCLEF photo annotation challenge [10]. The database is a subset of MIR FLICKR-1M image collection from thousands of the real-world users under a creative common license. The participants of the challenge were asked to elaborate methods in order to automatically annotate a test set of images with a collection of visual concepts. The task could be solved using three different types of approaches [10]:

- Visual: automatic annotation using visual information only.
- Textual: automatic annotation using textual information only (Flickr user tags and image metadata).
- Multimodal: automatic multimodal annotation using visual information and/or Flickr user tags and/or EXIF information.

The performance was quantitatively measured by the Mean interpolated Average Precision (MiAP) as the standard evaluation measure, while the example-based evaluation applies the example-based F-Measure (F-Ex) and Semantic R-Precision

{0A432C9F-1732-45E6-90F7-A6A7B75FA889}.jpg
Flickr user tags: peacock, bird, beautiful, pretty, feathers, waimea, waimeafalls, explore, animal, interestingness

Fig. 1.4 An example image with sparse Flickr user tags, including however semantic concepts ("bird," "beautiful," and "interestingness," etc.)

(SRPrecision) [10]. In the following experiments, we focus on the evaluation using MiAP.

The proposed approach was investigated under the following conditions: (1) the performance of SWLF using only visual features; (2) the performance of SWLF using only textual features; (3) the effect of combining textual and visual features through our SWLF scheme and the performance on the SWLF variants; (4) the performance of our runs at ImageCLEF 2011 and 2012 Flickr photo annotation challenge; (5) a discussion on the impact of the validation size on the generalization ability of SWLF. We start by describing the experimental setup.

1.5.1 Experimental Setup

The initial training dataset, provided by ImageClef 2011 for the photo annotation challenge, was first divided into a training set (50 %, 4005 images) and a validation set (50 %, 3995 images), and balanced the positive samples of most concepts as half for training and half for validation. An example of an image with its associated tags is given in Fig. 1.4. The proposed features, both textual and visual, were then extracted from the training and validation sets. Support Vector Machines (SVM) [60] were chosen as classifiers (or experts) for their effectiveness both in terms of computation complexity and classification accuracy. An SVM expert was trained for each concept and each type of features, as described in Sect. 1.4. Following Zhang et al. [61], we used χ^2 kernel for histogram-based features and RBF kernels for the other features. The RBF and χ^2 kernel functions are defined by the following equations:

$$K_{rbf}(F, F') = \exp^{-\frac{1}{2\sigma^2}\|(F-F')\|^2} \tag{1.7}$$

$$K_{\chi^2}(F, F') = \exp^{\frac{1}{I}\sum_{i=1}^{n}\frac{(F_i-F'_i)^2}{F_i+F'_i}} \tag{1.8}$$

where F and F' are the feature vectors, n is their size, I is the parameter for normalizing the distances which was set at the average value of the training set, and σ was set at $\sqrt{n/2}$.

We made use of the LibSVM library [62] as SVM implementation (C-Support Vector Classification). The tuning of the different parameters for each SVM expert was performed empirically according to our experiments, in which the weight of the samples from the negative class was set at 1, and the weight of the samples from the positive class was optimized on the validation set using a range of 1 through 30.

1.5.2 Results on the SWLF Using Visual Features

We performed the SWLF scheme for fusing visual features, and found that fusing the top five features yield the best MiAP (35.89 %) on the validation set, as shown in Fig. 1.5a. The results indicated that the weighted score and mean rules through SWLF outperforms the other two fusion rules, namely min and max, and the MiAP performance is increased by 3 % using the weighted score-based SWLF scheme compared to 32.9 % achieved by the best single visual feature (RGB-SIFT). As a result, the visual model, which we submitted to the photo annotation task at ImageCLEF 2011, performed the fusion of the top five best visual features using the score-based SWLF scheme. As shown in Fig. 1.5b, the fused experts proved to have a very good generalization skill on the test set. It can be seen that the weighted score and mean fusion methods perform better than the others, and the best fused experts on the validation set, which combine the top five features, achieved a MiAP of 35.54 % on the test set, in comparison with a MiAP of 35.89 % achieved by the same fused experts on the validation set.

As it can be seen from Fig. 1.5, the performance by score-based SWLF is not very different from the performance by mean-based SWLF even though the former performs slightly better than the latter, especially on the test set. The reason is that each expert is weighted by its normalized iAP, the weights computed by SWLF for all the experts are not very different, even roughly the same for the 10 best visual features (5 SIFT features and 5 LBP features). In the SWLF algorithm, each expert is weighted by its normalized iAP. This eventually results in roughly the same weights

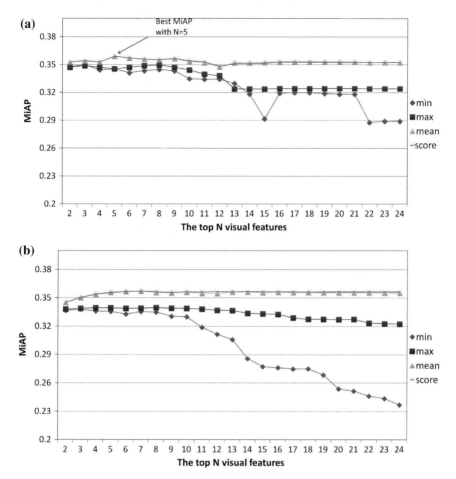

Fig. 1.5 The MiAP performance of different fusion methods based on SWLF scheme using the visual features on the validation set (**a**) and test set (**b**). As required by SWLF, the features are first sorted by descending order in terms of iAP of their corresponding experts. Then, the number of fused features N is increased from 1 to 24 (total number of visual features)

for the first 10 experts, and the range of weights which is not very big when N goes beyond 10, especially after weight normalization.

1.5.3 Results on the SWLF Using Text Features

We also applied the SWLF scheme to fuse textual features. The results as shown in Fig. 1.6a indicate that the combination of the top five best features yield the best MiAP value, and the weighted score-based SWLF scheme outperforms the other

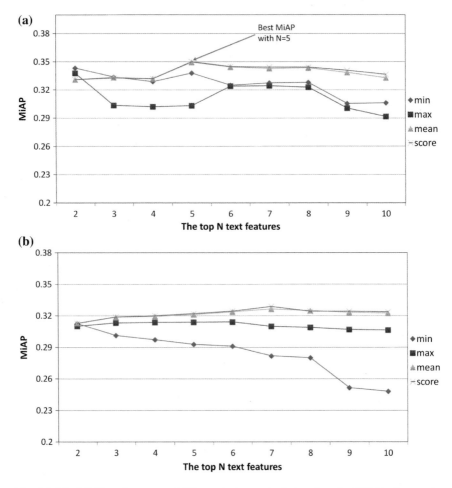

Fig. 1.6 The MiAP performance of different fusion methods based on the SWLF scheme using textual features on the validation set (**a**) and test set (**b**). As required by SWLF, the features are first sorted by descending order in terms of iAP of their corresponding experts. Then, the number of fused features N is increased from 1 to 10 (total number of textual features)

fusion rules, and achieves a MiAP of 35.01 % which improves by six points the MiAP of 29.1 % achieved by the best single textual feature (*txtf_99ps*) on the validation set. As a result, we implemented our text-based prediction model using the weighted score-based SWLF scheme to fuse the top five best textual features. As shown in Fig. 1.6b, the fused experts using the top five features achieve a MiAP of 32.12 % on the test set. It thus displays a very good generalization skill when this last figure is compared with the MiAP of 35.01 % achieved by the same fused experts on the validation set. Again, we also discovered that the score and mean-based SWLFs perform better than the others.

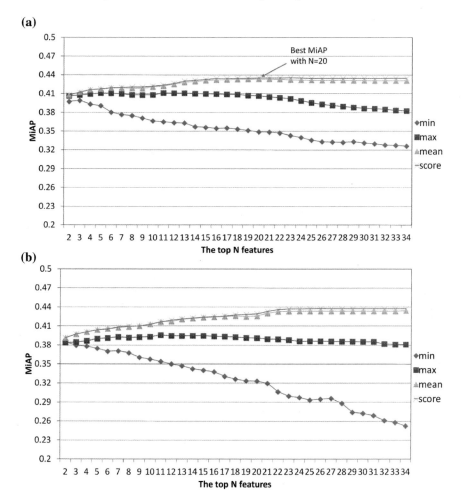

Fig. 1.7 The MiAP performance of SWLF_FN using different rules ("min," "max," "mean," and "score") for fusing visual and textual features using the validation set (**a**) and the test set (**b**)

1.5.4 Results on the SWLF and Its Variants Using All Features

Figure 1.7 presents the MiAP performance on all the features achieved by the SWLF scheme, using the "score" rule for combining experts which is compared with the standard fusion operators "min," "max," and "mean." These results are given on both the validation and test sets and show the evolution of the MiAP as N, the number of features to be fused, is increased from 1 to 34.

As we can see from Fig. 1.7a, the max and min-based SWLF_FN schemes tend to decrease the MiAP when the number of features to be fused, N, is successively increased from 1 to 34. On the contrary, the performance of weighted score and

Table 1.3 The MiAP obtained by SWLF_FN, SWLF_VN and SWLF_SFS on the validation and test sets

Method	MiAP on the validation set (%)	MiAP on the test set (%)
SWLF_FN(N=20)	43.55	42.71
SWLF_FN(N=22)	43.53	43.69
SWLF_VN	**44.51**	38.61
SWLF_SFS	44.03	**43.93**

mean-based SWLF_FN schemes keeps increasing until N reaches 20 and then stays stable. While close to each other, the weighted score-based SWLF_FN scheme performs slightly better than the mean-based SWLF_FN scheme. These results demonstrate that the weighted score-based SWLF scheme performs consistently better than the mean, max, and min-based fusion rules. Figure 1.7b presents the results obtained using the test set. We can observe that the results are very close to those obtained using the validation set, which proves the very good generalization skill of SWLF_FN, particularly when using "mean" and "score" fusion rules.

A comparison of the MiAP obtained by the three SWLF variants (SWLF_FN, SWLF_VN and SWLF_SFS) is provided in Table 1.3. It confirms the good generalization skill of SWLF since the MiAP obtained on the test set is very similar to the one obtained on the validation set. The best result is obtained by SWLF_SFS with a MiAP of 43.93 % on the test set, closely followed by SWLF_FN (with $N = 22$) with a MiAP of 43.69 %. SWLF_VN is the least efficient among SWLF variants. Indeed, although it performs slightly better than SWLF_FN and SWLF_SFS on the validation set, its performance drops by more than 5 % on the test set. This tends to suggest that SWLF_VN, in optimizing the iAP on a per class-basis, is more prone to overfitting than SWLF_FN and SWLF_SFS, thus leading to a more severe performance drop on unseen data (test dataset).

Figure 1.8 presents the iAP obtained by SWLF_FN, SWLF_VN and SWLF_SFS for each of the 99 concepts that had to be detected within the ImageCLEF 2011 Photo Annotation challenge. One can notice that the slight superiority of SWLF_SFS over SWLF_FN based on the global MiAP is respected for most of the concepts, as well as the lower results obtained by SWLF_VN. This figure also shows that some concepts are very well detected such as "Neutral_Illumination," "Outdoor," and "Sky" with an iAP around 90 % whereas some are very difficult to detect such as "Abstract," "Boring," and "Work" with an iAP lower than 10 %.

1.5.5 Results at ImageCLEF 2011 Photo Annotation Challenge

We submitted five runs to the ImageCLEF 2011 photo annotation challenge (two textual prediction models, one visual prediction model and two multimodal prediction models). All runs were evaluated on the test set composed of 10,000 images.

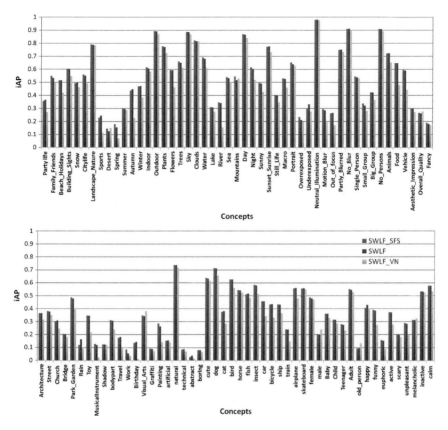

Fig. 1.8 The iAP obtained by SWLF_FN, SWLF_VN and SWLF_SFS for the 99 concepts of ImageCLEF 2011 Photo Annotation challenge

They were learnt by the weighted score-based SWLF on the training and validation sets using the features described in the previous sections, including 10 textual ones (using user tags) and 24 visual ones. The two textual prediction models made use of only textual features extracted from the user tags associated with an input image for predicting the visual concepts within it. The visual prediction model made use of only visual features while the two multimodal prediction models made joint use of the textual and visual features. We did not use the EXIF meda data provided for the photos.

1. **textual_model_1**: the combination of the top 4 features among the 10 textual features for each concept based on the weighted score SWFL scheme.
2. **multimodal_model_2**: the combination of the top 21 features among 34 visual and textual features for each concept based on the weighted score SWFL scheme.
3. **textual_model_3**: the combination of the top 5 features among the 10 textual features for each concept based on the weighted score SWFL scheme.

Table 1.4 The results of our runs submitted to ImageCLEF 2011 photo annotation challenge

Submitted runs	MiAP (%)	F-Ex (%)	SR-Precision (%)
textual_model_1	31.76	43.17	67.49
multimodal_model_2	42.96	57.57	71.74
textual_model_3	32.12	40.97	67.57
visual_model_4	35.54	53.94	72.50
multimodal_model_5	**43.69**	**56.69**	**71.82**
The best results:TUBFI [18]	44.34	56.59	55.86

4. **visual_model_4**: the combination of the top 5 features among the 24 visual features for each concept based on the weighted score SWFL scheme.
5. **multimodal_model_5**: the combination of the top 22 features among the 34 visual and textual features for each concept based on the weighted score SWFL scheme.

Thanks to the combination of the textual and visual features using our weighted score-based SWFL scheme, our fifth multimodal run achieved a MiAP of 43.69 % which was ranked the second performance out of 79 runs on the MiAP evaluation, as shown in Table 1.4. Indeed, our best visual model with 35.5 % was awarded the fifth in comparison to the best performance of 38.8 % in visual configuration. Our best textual model with 32.1 % was ranked the fourth performance while the best performance of textual modality was 34.6 %. Our weighted score-based SWLF fusion method again demonstrated its effectiveness, displaying a MiAP of 43.69 % which improves the MiAP of 35.54 % of our visual prediction model by roughly 8 % and even by 11 % the MiAP of 32.12 % of our best textual prediction model. It ranked closely to the first place obtained by TUBFI (TU Berlin and Fraunhofer FIRST) [18], who employed a non-sparse multiple kernel learning to combine kernels produced from various BoW features including a soft mapping of textual Bag-of-Words (BoW) and Markov random walks based on frequent Flickr user tags.

1.5.6 Results at ImageCLEF 2012 Photo Annotation Challenge

The ImageCLEF 2012 photo annotation task continues along the same lines as previous years. It provides 15,000 images for training and 10,000 images for testing, and has 94 concepts to be detected, where a few old concepts have been removed and a few new ones have been added. To assess the performance of the runs submitted by the teams, except the MiAP and F1 that have been used in ImageCLEF2011, the organizer applied the new evaluation measurement: Geometric Mean Average Precision (GMAP), which is an extension to MAP. When comparing runs with each other, the GMAP specifically highlights improvements obtained on relatively difficult concepts. Indeed, increasing the average precision of a concept from

0.05 to 0.10 has a larger impact in its contribution to the GMAP than increasing the average precision from 0.25 to 0.30. To compute the non-interpolated GMAP (GMnAP) and the interpolated GMAP (GMiAP), the same procedure is conducted the same as MnAP and MiAP, but instead with average the logs of the average precision for each concept. GMAP is obtained by exponentiating the resulting averages. Meanwhile, a very small epsilon value to each average precision is added before computing its log, so as to avoid taking the log of an average precision of zero. The epsilon value is very small and its effect on the final GMAP is negligible.

For this task, we continue to use the visual features described in Sect. 1.4.1. Moreover, we add one popular descriptor TOPSURF, and improve mid-level features harmony and dynamism using spatial pyramid strategy. We also include the distributional term representation DOR and DOR-TF/IDF [63] into the text features. In addition, we develop to use the color SIFT features based on soft assignment [64]. By using the SWLF fusion scheme, we have also submitted five runs (two text models, one visual model, and two multi-models), and the details of configuration can be found in [65].

In this task, 18 teams submitted in total 80 runs, of which 17 runs exclusively used textual features, 28 runs exclusively used visual features, and 35 runs used a multimodal approach. We present the overall evaluation results according to the MiAP, GMiAP, and micro-F1 in Table 1.5 to get an understanding of the best results irrespective of the features used, where in the Feature column the letter T refers to the textual configuration, V to the visual configuration, and M to the multimodal configuration. In the tables, the ranks indicate the position at which the best run appeared in the results. To compare only runs using the same configuration we present separate results for the textual features in Table 1.6, the visual features in Table 1.7.

From the overall evaluation results shown in Table 1.5, we can see that one of our multimodal results achieved a MiAP of 43.67 % which ranks first out of 80 runs. It is confirmed once again that our multimodal approaches, by fusing with the efficient textual and visual features using our weighted score-based SWFL scheme, are effective for the generic VCR task. Moreover, our results not only achieve the best performance according to MiAP measurement, but also in terms of the two other measures, namely micro-F1 and GMnAP. The performances of our visual and textual modalities are also significant since they obtained the first rank except for the visual configuration using the micro-F1 measure. This indicates that our proposed fusion method SWLF also works well for combining homogeneous information sources such as the visual or textual modalities.

1.5.7 Discussion on the Generalization Skill of SWLF

In our experimental setup, defined in Sect. 1.5.1, the initial training dataset was divided into roughly two equal parts: a training set and a validation set. The SWLF uses the training set to train an ensemble of experts (classifiers), one for each visual

Table 1.5 Summary of the annotation results for the evaluation per concept and image for the best overall run per team per evaluation measure

Team	Rank	MiAP	Feature	Team	Rank	GMiAP	Feature	Team	Rank	Micro-F1	Feature
LIRIS	1	0.4367	M	LIRIS	1	0.3877	M	LIRIS	1	0.5766	M
DMS-SZTAKI	3	0.4258	M	DMS-SZTAKI	3	0.3676	M	DMS-SZTAKI	3	0.5731	M
CEA LIST	6	0.4159	M	CEA LIST	5	0.3615	M	NII	6	0.5600	V
ISI	7	0.4136	M	ISI	7	0.3580	M	ISI	7	0.5597	M
NPDPILIP6	16	0.3437	V	NPDPILIP6	16	0.2815	V	MLKD	16	0.5534	V
NII	22	0.3318	V	NII	21	0.2703	V	CEA LIST	20	0.5404	M
CERTH	28	0.3210	M	MLKD	28	0.2567	V	CERTH	26	0.4950	M
MLKD	29	0.3185	V	CERTH	29	0.2547	M	IMU	30	0.4685	T
IMU	36	0.2441	T	IMU	35	0.1917	T	KIDS NUTN	34	0.4406	M
UAIC	38	0.2359	V	UAIC	39	0.1685	V	UAIC	35	0.4359	V
MSATL	41	0.2209	T	MSATL	42	0.1653	T	NPDPILIP6	37	0.4228	V
IL	46	0.1724	T	IL	45	0.1140	T	IL	49	0.3532	T
KIDS NUTN	47	0.1717	M	KIDS NUTN	49	0.0984	M	URJCyUNED	50	0.3527	T
BUAA AUDR	52	0.1423	V	BUAA AUDR	51	0.0818	V	PRA	54	0.3331	V
UNED	55	0.1020	V	UNED	55	0.0512	V	MSATL	57	0.2635	T
DBRIS	58	0.0976	V	DBRIS	57	0.0476	V	BUAA AUDR	58	0.2592	M
PRA	65	0.0900	V	PRA	66	0.0437	V	UNED	66	0.1360	V
URJCyUNED	77	0.0622	V	URJCyUNED	77	0.0254	V	DBRIS	69	0.1070	V

Table 1.6 Summary of the annotation results for the evaluation per concept and image for the best textual run per team per evaluation measure

Team	Rank	MiAP	Team	Rank	GMiAP	Team	Rank	Micro-F1
LIRIS	1	0.3338	LIRIS	1	0.2771	LIRIS	1	0.4691
CEA LIST	3	0.3314	CEA LIST	2	0.2759	IMU	2	0.4685
IMU	4	0.2441	IMU	4	0.1917	CEA LIST	5	0.4452
CERTH	6	0.2311	CERTH	7	0.1669	MLKD	7	0.3951
MSATL	8	0.2209	MSATL	9	0.1653	CERTH	8	0.3946
IL	11	0.1724	IL	11	0.1140	IL	10	0.3532
BUAA AUDR	13	0.1423	BUAA AUDR	13	0.0818	URJCyUNED	11	0.3527
UNED	14	0.0758	UNED	14	0.0383	MSATL	13	0.2635
MLKD	15	0.0744	MLKD	15	0.0327	BUAA AUDR	14	0.2167
URJCyUNED	17	0.0622	URJCyUNED	17	0.0254	UNED	16	0.0864

Table 1.7 Summary of the annotation results for the evaluation per concept and image for the best visual run per team per evaluation measure

Team	Rank	MiAP	Team	Rank	GMiAP	Team	Rank	Micro-F1
LIRIS	1	0.3481	LIRIS	1	0.2858	NII	1	0.5600
NPDILIP6	2	0.3437	NPDILIP6	2	0.2815	MLKD	6	0.5534
NII	6	0.3318	NII	5	0.2703	ISI	7	0.5451
ISI	10	0.3243	ISI	10	0.2590	LIRIS	8	0.5437
MLKD	11	0.3185	MLKD	11	0.2567	CERTH	9	0.4838
CERTH	13	0.2628	CERTH	13	0.1904	UAIC	10	0.4359
UAIC	14	0.2359	UAIC	14	0.1685	NPDILIP6	11	0.4228
UNED	15	0.1020	UNED	15	0.0512	PRA	15	0.3331
DBRIS	16	0.0976	DBRIS	16	0.0476	URJCyUNED	18	0.1984
PRA	22	0.0873	PRA	23	0.0437	UNED	19	0.1360
MSATL	24	0.0868	MSATL	25	0.0414	DBRIS	22	0.1070
URJCyUNED	28	0.0622	URJCyUNED	28	0.0254	MSATL	23	0.1069

concept and each type of features. It then selects and combines the best experts while optimizing the overall MiAP on the validation set. In the following, we give an analysis on the generalization ability of SWLF on the overall performance.

The first question concerns the generalization skill of a fused expert through SWLF on unseen data. In Sects. 1.5.2, 1.5.3 and 1.5.4, we already depicted the good generalization skill of the fused experts through the weighted score-based SWLF, on test dataset. Table 1.8 further highlights such a behavior of the fused experts in displaying their MiAP performance both on the validation and test dataset. The prediction models in bold correspond to the best prediction model learned through SWLF on the validation set. It can be seen that the best fused experts learned on the validation set keeps a quite good generalization skill as the performance only drops slightly on the test set. In our submission, we anticipated this performance drop in particular for multimodal prediction models. Instead of submitting the best multimodal model

Table 1.8 MiAP performance comparison of the fused experts learned through the weighted score-based SWLF on the validation set versus the test set

Prediction model	Nb of fused experts N	Validation set	Test set
textual_model_1	4	33.21	31.76
textual_model_3	5	35.01	32.12
visual_model_4	5	35.89	35.54
multimodal_model	20	43.54	42.71
multimodal_model_2	21	43.52	42.96
multimodal_model_5	22	43.53	43.69

The prediction models in bold correspond to the best fused experts learnt through weighted score-based SWLF on the validation set

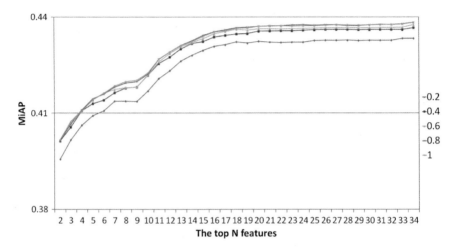

Fig. 1.9 The MiAP performance on the test dataset of the fused experts through SWLF when varying the size of the validation dataset from 20 to 100 % of the size of the original validation set

on the validation set which combines the best 20 features, we submitted two multimodal runs, namely *multimodal_model_4* and *multimodal_model_5*, making use of 21 and 22 best features, respectively. Surprisingly enough, our best multimodal run, *multimodal_model_5*, which was ranked the second best MiAP performance out of 79 runs, proves to perform slightly better on the test set than on the validation set.

The second question is how the size of the validation set impacts the generalization ability of a fused expert learned through SWLF. For this purpose, we evaluated, as shown in Fig. 1.9, the performance of the fused multimodal experts learnt through the score-weighted SWLF on the validation set, by varying the size of that validation set. The results on the test set were achieved by varying the size of the validation set from 20 to 100 % of the size of the original validation set, i.e., 3,995 images as specified in Sect. 1.5.1, while keeping the training set unchanged. The x axis displays the number of fused experts while the y axis gives the MiAP performance. The curves in different colors plot the MiAP performance using different size of the validation set. From this figure, we can see that the SWLF performance keeps increasing with the size of the validation set, and the improvement becomes slight from 40 % of the size of the original validation set. Given the size of a validation set, the fused expert displays a similar behavior: the performance increases quickly when N varies from 1 to 20, then it subsequently remains stable.

1.6 Conclusion

We have presented in this chapter a novel Selective Weighted Late Fusion (SWLF) that iteratively selects the best features and weights the corresponding scores for each concept at hand to be classified. Three variants of SWLF, namely SWLF_FN, SWLF_VN and SWLF_SFS, have been proposed and compared.

Experiments were conducted on the image collection within the ImageCLEF Photo Annotation challenge. In 2011, our submission using SWLF_FN obtained a MiAP of 43.69 % for the detection of the 99 visual concepts that ranked second out of the 79 submitted runs. In 2012, a MiAP of 43.67 % was obtained that ranked first out of the 80 submitted runs. The experimental results have also shown that SWLF, in efficiently fusing visual and textual features, display a very good generalization ability on unseen data for the image annotation task with a multi-label scenario.

In our future work, we plan to investigate other directions for learning high-level features using machine learning methods. Indeed, many successful object recognition approaches are proposed based on local features such as SIFT and HOG. However, these only capture low-level edge information and it has proven difficult to design features that effectively capture mid-level cues (such as edge intersections) or high-level representation (such as object parts). Recent developments in machine learning, known as "Deep Learning," have shown how hierarchies of features can be learned directly from data. These machine learning approaches, as applied to learning high-level features for visual concept recognition in images and video, could further improve the performances by increasing the discriminative ability of low-level features for Visual Concept Recognition.

References

1. Salton G, McGill MJ (1986) Introduction to modern information retrieval. McGraw-Hill Inc, New York
2. Smeulders AWM, Worring M, Santini S, Gupta A, Jain R (2000) Content-based image retrieval at the end of the early years. IEEE Trans Pattern Anal Mach Intell 22:1349–1380
3. Picard RW (2000) Affective computing. MIT press, Cambridge
4. Mojsilović A, Gomes J, Rogowitz B (2004) Semantic-friendly indexing and quering of images based on the extraction of the objective semantic cues. Int J Comput Vision 56:79–107
5. Snelick R, Uludag U, Mink A, Indovina M, Jain A (2005) Large-scale evaluation of multimodal biometric authentication using state-of-the-art systems. IEEE Trans Pattern Anal Mach Intell 27:450–455

6. Machajdik J, Hanbury A (2010) Affective image classification using features inspired by psychology and art theory. In: Proceedings of the international conference on Multimedia, ACM, pp 83–92
7. Liu N, Dellandréa E, Chen L, Zhu C, Zhang Y, Bichot CE, Bres S, Tellez B (2013) Multimodal recognition of visual concepts using histograms of textual concepts and selective weighted late fusion scheme. Comput Vis Image Underst 117:493–512
8. Everingham M, Van Gool L, Williams CK, Winn J, Zisserman A (2010) The pascal visual object classes (voc) challenge. Int J Comput Vis 88:303–338
9. Smeaton AF, Over P, Kraaij W (2006) Evaluation campaigns and trecvid. In: MIR '06: Proceedings of the 8th ACM international workshop on multimedia, information retrieval, pp 321–330
10. Nowak S, Nagel K, Liebetrau J (2011) The clef 2011 photo annotation and concept-based retrieval tasks. In: CLEF workshop notebook paper
11. Guillaumin M, Verbeek JJ, Schmid C (2010) Multimodal semi-supervised learning for image classification. In: Proceedings of CVPR, pp 902–909
12. Snoek CGM, Worring M, Smeulders AWM (2005) Early versus late fusion in semantic video analysis. In: Proceedings of the 13th annual ACM international conference on multimedia, pp 399–402
13. Ah-Pine J, Bressan M, Clinchant S, Csurka G, Hoppenot Y, Renders JM (2009) Crossing textual and visual content in different application scenarios. Multimedia Tools Appl 42:31–56
14. Snoek CGM, Worring M, Geusebroek JM, Koelma DC, Seinstra FJ (2004) The mediamill trecvid 2004 semantic video search engine. In: Proceedings of the TRECVID workshop
15. Westerveld T, Vries APD, van Ballegooij A, de Jong F, Hiemstra D (2003) A probabilistic multimedia retrieval model and its evaluation. EURASIP J Appl Signal Process 2003:186–198
16. Noble WS et al (2004) Support vector machine applications in computational biology. In: Schoelkopf B, Tsuda K, Vert, J-P (eds) Kernel methods in computational biology. MIT Press, Cambridge, pp 71–92
17. Pinquier J, Karaman S, Letoupin L, Guyot P, Mégret R., Benois-Pineau J, Gaestel Y, Dartigues JF (2012) Strategies for multiple feature fusion with hierarchical hmm: application to activity recognition from wearable audiovisual sensors. In: Proceedings of 21st international conference on pattern recognition (ICPR), IEEE, pp 3192–3195
18. Binder A, Samek W, Kloft M, Müller C, Müller KR., Kawanabe M (2011) The joint submission of the tu berlin and fraunhofer first (tubfi) to the imageclef2011 photo annotation task. In: CLEF workshop notebook paper
19. Nagel K, Nowak S, Kühhirt U, Wolter K (2011) The Fraunhofer IDMT at ImageCLEF 2011 photo annotation task. In: Proceedings of CLEF (Notebook Papers/Labs/Workshop)
20. Csurka G, Dance C, Fan L, Willamowski J, Bray C (2004) Visual categorization with bags of keypoints. In: Workshop on statistical learning in computer vision, ECCV, vol 1, p 22
21. Quenot G, Benois-Pineau J, Mansencal B, Rossi E, Cord M, Precioso F, Gorisse D, Lambert P, Augereau B, Granjon L et al (2008) Rushes summarization by IRIM consortium: redundancy removal and multi-feature fusion. In: Proceedings of the 2nd ACM TRECVid video summarization workshop, ACM, pp 80–84
22. Wu Y, Chang EY, Chang KCC, Smith JR (2004) Optimal multimodal fusion for multimedia data analysis. In: Proceedings of the 12th annual ACM international conference on Multimedia, pp 572–579
23. Znaidia A, Borgne HL, Popescu A (2011) CEA list's participation to visual concept detection task of ImageCLEF 2011. In: CLEF workshop notebook paper
24. Kittler J, Hatef M, Duin RPW, Matas J (1998) On combining classifiers. IEEE Trans Pattern Anal Mach Intell 20:226–239
25. Ben Soltana W, Huang D, Ardabilian M, Chen L, Ben Amar C (2010) Comparison of 2D/3D features and their adaptive score level fusion for 3D face recognition. In: 3D data processing, visualization and transmission (3DPVT)
26. Pudil P, Novovičová J, Kittler J (1994) Floating search methods in feature selection. Pattern Recogn Lett 15:1119–1125

27. Rokach L (2010) Ensemble-based classifiers. Artif Intell Rev 33:1–39
28. Breiman L, Breiman L (1996) Bagging predictors. Mach Learn 24:123–140
29. Fergus R, Fei-Fei L, Perona P, Zisserman A (2005) Learning object categories from google's image search. In: 10th IEEE international conference on computer vision ICCV, IEEE, vol 2, pp 1816–1823
30. Schroff F, Criminisi A, Zisserman A (2007) Harvesting image databases from the web. In: IEEE 11th international conference on computer vision, ICCV, pp 1–8
31. Wang G, Hoiem D, Forsyth DA (2009) Building text features for object image classification. In: Proceedings of CVPR, pp 1367–1374
32. Ojala T, Pietikäinen M, Harwood D (1996) A comparative study of texture measures with classification based on featured distributions. Pattern Recogn 29:51–59
33. Zhu C, Bichot CE, Chen L (2010) Multi-scale color local binary patterns for visual object classes recognition. In: Proceedings of ICPR, pp 3065–3068
34. Pujol A, Chen L (2007) Line segment based edge feature using hough transform. In: Proceedings of the 7th IASTED international conference on visualization, imaging and image processing, ACTA Press, pp 201–206
35. Lowe DG (2004) Distinctive image features from scale-invariant keypoints. Int J Comput Vis 60:91–110
36. Mikolajczyk K, Schmid C (2004) Scale and affine invariant interest point detectors. Int J Comput Vis 60:63–86
37. Lazebnik S, Schmid C, Ponce J (2006) Beyond bags of features: spatial pyramid matching for recognizing natural scene categories. In: Proceedings of CVPR, vol 2, pp 2169–2178
38. Li FF, Perona P (2005) A Bayesian hierarchical model for learning natural scene categories. In: Proceedings of CVPR, vol 2, pp 524–531
39. Van de Sande KEA, Gevers T, Snoek CGM (2010) Evaluating color descriptors for object and scene recognition. IEEE Trans Pattern Anal Mach Intell 32:1582–1596
40. Tola E, Lepetit V, Fua P (2010) Daisy: an efficient dense descriptor applied to wide-baseline stereo. IEEE Trans Pattern Anal Mach Intell 32:815–830
41. Zhu C, Bichot CE, Chen L (2011) Visual object recognition using daisy descriptor. In: Proceedings of ICME, pp 1–6
42. Dunker P, Nowak S, Begau A, Lanz C (2008) Content-based mood classification for photos and music: a generic multi-modal classification framework and evaluation approach. In: Proceedings of multimedia information retrieval, pp 97–104
43. Liu N, Dellandréa E, Tellez B, Chen L (2011) Evaluation of features and combination approaches for the classification of emotional semantics in images. In: International conference on computer vision, theory and applications (VISAPP)
44. Liu N, Dellandréa E, Tellez B, Chen L, Chen L (2011) Associating textual features with visual ones to improve affective image classification. In: Proceedings of ACII, vol 1, pp 195–204
45. Valdez P, Mehrabian A (1994) Effects of color on emotions. J Exp Psychol Gen 123:394–409
46. Itten J, Van Haagen E (1973) The art of color: the subjective experience and objective rationale of color. Van Nostrand Reinhold, New York
47. Tamura H, Mori S, Yamawaki T (1978) Texture features corresponding to visual perception. IEEE Trans Syst Man Cybern 8(6):460–472
48. Haralick RM (1979) Statistical and structural approaches to texture. Proc IEEE 67:786–804
49. Anstey NA (1966) Correlation techniques—a reivew. Can J Explor Geophys 2:55–82
50. van de Sande K. Colordescriptor software. http://www.colordescriptors.com
51. Colombo C, Bimbo AD, Pala P (1999) Semantics in visual information retrieval. IEEE Multimedia 6:38–53
52. Dellandréa E, Liu N, Chen L (2010) Classification of affective semantics in images based on discrete and dimensional models of emotions. In: International workshop on content-based multimedia indexing (CBMI), pp 99–104
53. Ke Y, Tang X, Jing F (2006) The design of high-level features for photo quality assessment. In: IEEE computer society conference on computer vision and pattern recognition, vol 1, pp 419–426

54. Datta R, Li J, Wang JZ (2005) Content-based image retrieval: approaches and trends of the new age. In: Proceedings on multimedia information retrieval, pp 253–262
55. Viola PA, Jones MJ (2001) Robust real-time face detection. In: Proceedings of CCV, vol 57, pp 137–154
56. Budanitsky A, Hirst G (2001) Semantic distance in wordnet: an experimental, application-oriented evaluation of five measures. In: Workshop on WordNet and other lexical resources, 2nd meeting of the North American chapter of the association for computational linguistics
57. Miller GA (1995) Wordnet: a lexical database for English. Commun ACM 38:39–41
58. Huiskes MJ, Lew MS (2008) The MIR flickr retrieval evaluation. In: Proceedings on multimedia information retrieval, pp 39–43
59. Huiskes MJ, Thomee B, Lew MS (2010) New trends and ideas in visual concept detection: the MIR flickr retrieval evaluation initiative. In: MIR '10: Proceedings of the 2010 ACM international conference on multimedia, information retrieval, pp 527–536
60. Vapnik VN (1995) The nature of statistical learning theory. Springer New York Inc., New York
61. Zhang J, Marszaek M, Lazebnik S, Schmid C (2007) Local features and kernels for classification of texture and object categories: a comprehensive study. Int J Comput Vis 73:213–238
62. Chang CC, Lin CJ (2011) LIBSVM: a library for support vector machines. ACM Trans Intell Syst Technol 2:1–27
63. Escalante HJ, Montes M, Sucar E (2011) Multimodal indexing based on semantic cohesion for image retrieval. Inf Retrieval 15:1–32
64. van Gemert JC, Veenman CJ, Smeulders AWM, Geusebroek JM (2010) Visual word ambiguity. IEEE Trans Pattern Anal Mach Intell 32:1271–1283
65. Liu N, Zhang Y, Dellandréa E, Bres S, Chen L (2012) LIRIS-Imagine at ImageCLEF 2012 photo annotation task. In: CLEF workshop notebook paper

Chapter 2
Bag-of-Words Image Representation: Key Ideas and Further Insight

Marc T. Law, Nicolas Thome and Matthieu Cord

Abstract In the context of object and scene recognition, state-of-the-art performances are obtained with visual Bag-of-Words (BoW) models of mid-level representations computed from dense sampled local descriptors (e.g., Scale-Invariant Feature Transform (SIFT)). Several methods to combine low-level features and to set mid-level parameters have been evaluated recently for image classification. In this chapter, we study in detail the different components of the BoW model in the context of image classification. Particularly, we focus on the coding and pooling steps and investigate the impact of the main parameters of the BoW pipeline. We show that an adequate combination of several low (sampling rate, multiscale) and mid-level (codebook size, normalization) parameters is decisive to reach good performances. Based on this analysis, we propose a merging scheme that exploits the specificities of edge-based descriptors. Low and high contrast regions are pooled separately and combined to provide a powerful representation of images. We study the impact on classification performance of the contrast threshold that determines whether a SIFT descriptor corresponds to a low contrast region or a high contrast region. Successful experiments are provided on the Caltech-101 and Scene-15 datasets.

M. T. Law (✉) · N. Thome · M. Cord
LIP6, UPMC—Sorbonne University, Paris, France
e-mail: Marc.Law@lip6.fr

N. Thome
e-mail: Nicolas.Thome@lip6.fr

M. Cord
e-mail: Matthieu.Cord@lip6.fr

2.1 Introduction

Image classification refers to the ability of predicting a semantic concept based on the visual content of the image. This topic is extensively studied due to its large number of applications in areas as diverse as Image Processing, Information Retrieval, Computer Vision, and Artificial Intelligence [10]. This is one of the most challenging problems in these domains. For instance, in the context of Computer Vision, the ability to predict complex semantic categories, such as scenes or objects, from the pixel level, is still a very hard task. How to properly represent images for successfully categorizing images, i.e., filling the semantic gap, remains a major issue for computer vision researchers.

Different methodologies have been explored in the last decade to fulfill this goal. Biologically inspired models [37, 42] try to mimic the mammalian visual system, and show interesting performances for classification and detection. Recently, deep learning has attracted lots of attention due to the large success of deep convolutional nets in the Large Scale Visual Recognition Challenge 2012 (ILSVRC2012).[1] Using pixels as input, the network automatically learns useful image representations for the classification task. The results reveal that deep learning significantly outperforms state-of-the-art computer vision representations competitors [29]. However, although this trend is unquestionable for this large-scale context (1 million training examples), the feasibility of reaching state-of-the-art performances in other complex datasets with fewer training examples remains unclear.

Therefore, the Bag-of-words (BoW) model [39] that proved to be the leading strategy in the last decade, remains a very competitive representation. Two main breakthroughs have been reached that explain the BoW model strength. The first one is the design of discriminative low-level local features, such as Scale-Invariant Feature Transform (SIFT) [32] and Histograms of oriented Gradients (HoG) [12]. The second one is the emergence of mid-level representations inspired from the text retrieval community. Indeed, coding local features and aggregating the codes (pooling) make it possible to output a vectorial representation for each image. Subsequently, this representation can be used to train powerful statistical learning models, e.g., Support Vector Machine (SVM) [11], or to model visual attention maps by feature weighting and selection using biologically inspired methods [23, 38, 47].

Extensive studies have been carried out for adapting to images, the initial method [39] inspired by text information retrieval. In particular, many attempts for improving the coding and pooling steps have been done. In this chapter, we first investigate the BoW pipeline in terms of parameter setting and feature combination for classification. We do believe that such an analysis should help to clarify the real difference between mid-level representations for a classification purpose. Based on this study, we also introduce an early fusion [41] method that takes into account and distinguishes low contrast regions from high contrast regions in images. Low contrast regions are usually either completely removed and ignored from the mid-level representation of images, or processed as any common feature. The idea is to

---

[1] http://www.image-net.org/challenges/LSVRC/2012/

exploit occurrence statistics of low contrast regions and combine them with classical recognition methods applied on high contrast regions. The fusion we propose does not exploit low-level features of different natures (such as combining edge-based, color, metadata descriptors, etc.) but processes low-level features differently with regard to their gradient magnitude. We focus our experiments on the Caltech-101 [18] and Scene-15 [30] datasets, where most of state-of-the-art methods improving over the BoW model have been evaluated.

The remainder of the chapter is organized as follows: Section 2.2 gives an overview of the most significant mid-level BoW improvements, and clarifies the impact of other low-level and mid-level parameters on classification performances. Section 2.3 presents the classification pipeline evaluated in this chapter. In Sect. 2.4, we specifically study the pooling fusion of low and high contrast regions. With local edge-based descriptors (e.g., SIFT), the feature normalization process is likely to produce noisy features: we analyze the use of a thresholding procedure used in VLFEAT [44] to overcome this problem. In addition, we propose novel coding and pooling methods that are well adapted for handling low contrast regions. Section 2.5 provides a systematic evaluation of the impact on classification performances of the different parameters studied in the chapter.

2.2 Bag-of-Words Literature

In the BoW model, converting the set of local descriptors into the final image representation is performed by a succession of two steps: coding and pooling. In the original BoW model, coding consists of hard assigning each local descriptor to the closest visual word, while pooling averages the local descriptor projections. The final BoW vector can thus be regarded as a histogram counting the occurrences of each visual word in the image. Since the notion of 'word' is not as clear for image classification as for text retrieval, many efforts have been recently devoted to improve coding and pooling [4]. It is worth mentioning that extensive works have also been proposed to combine multiple low-level features. For example, Vedaldi et al. [45] and Gehler et al. [20] report that the performances can be significantly improved, using Multiple Kernel Learning (MKL) [2] or LP-boosting, respectively. Combining multiple low-level features is a complementary approach to the improvement of mid-level representation, and we do not give more details on such methods. Since this low-level combination is known to have a large impact on performances (for example in [20] the best reported results are above 82% while the best performing low-level feature is below 70%), the remainder of the methods studied by now focus on mono feature methods, where mid-level representations are extracted from a single low-level visual modality.

When dealing with images and visual codebooks, the hard assignment strategy induces an approximation of the local feature. To attenuate this quantization loss, soft-assignment attempts to smoothly distribute features to the codewords [21, 31]. In sparse coding approaches [5, 48, 49], there is an explicit minimization of the feature reconstruction error, along with a prior regularization that encourages sparse solutions. However, one main drawback in sparse coding is that the code optimization needs to be solved for each descriptor. This makes inference very slow, especially when there are many descriptors or when the dictionary is large. As sparse coding are decoder networks, some approaches propose to learn encoding-decoder networks [22, 28], in which an encoder is concurrently learned to avoid performing the heavy sparse coding minimization. Another way to make coding more accurate is to have a vectorial coding scheme. In aggregate methods, such as Fisher vectors [34, 35], VLAD [26] or super-vectors [51], the difference in the feature space between the local descriptor and each codeword is stored. Despite their very good performances, these aggregate methods cause a huge inflation in the representation size, where a dimensionality of 1 million is common.

Regarding pooling, alternative strategies to averaging the codes have been studied. Max pooling is a promising alternative to sum pooling [5, 7, 31, 48, 49], especially when linear classifiers are used. Other works study the pooling beyond a scalar pooling. In [1, 16], the probability density function (pdf) of the distance between the local features and each codeword is estimated, providing a richer statistics of the codes than using average or max pooling (that output a scalar value). This vectorial pooling strategy is shown to improve performances in various image databases.

Finally, one important limitation of the visual BoW model is the lack of spatial information. The most popular extension to overcome this problem is the Spatial Pyramid Matching Scheme (SPM) [30]. SPM independently pools information from different images regions defined by a multi-resolution spatial grid and concatenates the histograms to form the final image vector. Despite its simplicity and rigidity, SPM generally brings a substantial gain in classification performances in most databases. Using more sophisticated models to incorporate spatial information in the vectorial representation, generally fails to improve performances over SPM. A noticeable exception is [13], where a graph-matching kernel strategy is used to model spatial alignment between images. The performance increases above 80 % which is outstanding for a mono feature method. However, the results were less impressive in a scene database such as Scene-15 [30]. Karaman et al. [27] propose a multi-layer structural approach for the task of object based image retrieval. The structural features are nested multi-layered local graphs built upon sets of SURF [3] feature points with Delaunay triangulation. A BoW framework is applied on these graphs. The multi-layer nature of the descriptors consists in scaling from trivial Delaunay graphs by increasing the number of nodes layer by layer up to graphs with maximal number of nodes. For each layer of graphs, its own visual dictionary is built. Finally, an interesting attempt including both spatial pooling and aggregation in the feature space is the work of Feng et al. [19]. They propose to learn both aspects of the pooling from data using a supervised criterion. Specifically, a per-class ℓ_p geometric pooling is introduced that learns the optimal pooling in between max and average pooling. A spatial weighting is also learned from data to maximize performances. They report the score of 82.6 %, which is, to the best of our knowledge, the state-of-the-art result using mono features in the Caltech-101 [18] database.

2 Bag-of-Words Image Representation

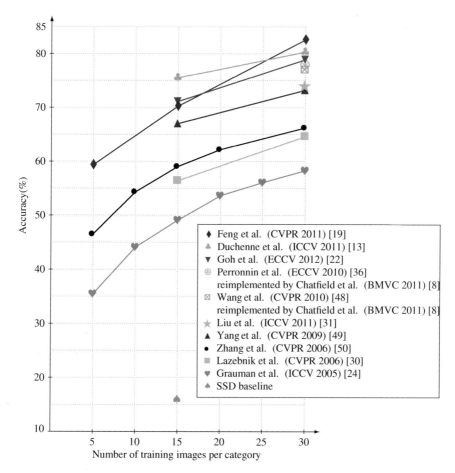

Fig. 2.1 State-of-the-art results since 2006 on the Caltech 101 for BoW pipeline methods in mono feature setup

Figure 2.1 shows the performance evolution of different mid-level representations, in the Caltech-101 database, using mono features. It is obvious that the mid-level steps improvement since 2006 significantly boosted performances: for example, using 30 training examples, there is a substantial gain of about 20 pt from the baseline SPM work of Lazebnik et al. [30] (∼64% in 2006) to the pooling learning method of Feng et al. [19] (∼83% in 2011). Nonetheless, the absolute numbers reported in the different publications to illustrate the improvement brought out by some mid-level representation sometimes also include differences in the feature computation or learning algorithms. These variations make the merits of a given mid-level representation confusing. This aspect has recently been studied by Chatfield et al. [8]. The authors re-implemented some of the most powerful mid-level representations (e.g., [35, 48, 51]), and provide an experimental comparison in

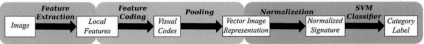

Fig. 2.2 BoW pipeline for classification

the PASCAL VOC [14] dataset. Apart from the fact that some methods are nonreproducible,[2] the main conclusion is that mid-level representation performances are strongly impacted by low-level feature extraction parameters. In particular, it is clear that mid-level representations benefit from a "heavy" low-level feature computation setup, mainly using a dense and multiscale extraction scheme.

In this chapter, we analyze the impact on classification performance of different parameters in the BoW pipeline. We extend here the analysis of Chatfield et al. [8] by including novel parameters, such as normalization of the image signature and a specific coding/pooling strategy for low contrast regions, that strongly impact performances. We focus our experiments on the Caltech-101 [18] and Scene-15 [30] databases, where most state-of-the-art methods improving over the BoW model have been evaluated.

2.3 Classification Pipeline

Figure 2.2 illustrates the whole classification pipeline studied in this chapter. Local features are first extracted in the input image, and encoded into an off-line trained dictionary. The codes are then pooled to generate the image signature. This mid-level representation is ultimately normalized before training the classifier. Each block of the figure is detailed in the following sections. In particular, we specify the main parameters of the BoW pipeline that have a strong impact on classification performances, and evaluate these parameters in Sect. 2.5.

2.3.1 Low-Level Feature Extraction

The first step of the BoW framework corresponds to local feature extraction. To extract local descriptors, one first issue is to detect relevant image regions. Many attempts have been done to achieve that goal, generally based on a saliency criterion:

---

[2] Chatfield et al. [8] report that their re-implementation of Zhou et al. [51] performs 6 % below the published results. From personal communication with the authors of Zhou et al. [51], the results reported in Chatfield et al. [8] are representative of the method performances, without including non trivial modifications not discussed in the chapter.

Harris detector [25] or its multiscale version [33], SIFT detector [32], etc. However, for classification tasks, most evaluations reveal that a regular grid-based sampling strategy leads to optimal performances [17]. Therefore, we follow this brute-force region selection scheme. In each patch, SIFT features [32] are computed because of their excellent performances attested in various datasets. SIFT features [32] have initially been designed for an image matching purpose. In this context, the ability to match image regions under various geometric and photometric deformations is crucial. For that reason, the SIFT descriptor is made rotationally invariant, by computing the gradient orientation relatively to the dominant orientation of the patch. However, in an image classification context, it is shown that ignoring the rotation invariance favorably impacts performances [43]. This is due to the fact that the orientation of the patch is actually informative for scene and object recognition, and we disable the orientation invariance in our experiments.

In the sampling process, two parameters have a strong impact on classification performances:

- **Sampling density** As we verify experimentally, the denser the sampling is, the better the performances get. The density is set through the spatial stride parameter and corresponds to the distance between the center of two closest extracted patches. In most recent published papers [5, 31, 48, 49], a commonly reported setup is to use a dense monoscale SIFT extraction scheme with a spatial stride set to 8 pixels. However, some authors provide publicly available code using different setups than those reported in their paper.[3]
- **Monoscale versus multiscale features** It is known [8] that using multiscale features increases the amount of low-level information for generating the mid-level signatures, and thus favorably impacts performances. Again, most approaches have been evaluated with a monoscale dense sampling strategy. Wang et al. [48] evaluate their method (LLC) in a multiscale setting, making the comparison with respect to other methods that use monoscale features somehow unfair.

2.3.2 Mid-Level Coding and Pooling Scheme

Let $\mathbf{X} = (\mathbf{x}_1, \ldots, \mathbf{x}_j, \ldots, \mathbf{x}_N)$ be the set of local descriptors in an image, where N is the number of local descriptors in the image. In the BoW model, the mid-level signature generation first requires a set of codewords $\left(\mathbf{b}_i \in \mathbb{R}^d\right)_{i=1}^M$ (d is the local descriptor's dimensionality, and M is the number of codewords). This set of codewords is called visual codebook or dictionary. Different strategies to compute the codebook exist. The codebook can be performed with a *static* clustering, e.g., Smith

---

[3] In the provided source codes for evaluation, the sampling is sometimes set to lower values: e.g., 6 pixels in http://www.ifp.illinois.edu/~jyang29/ScSPM.htm for Liu et al. [31] or http://users.cecs.anu.edu.au/~lingqiao/ for Liu et al. [31]. Compared to the value of 8 pixels, the performances decrease of about 1–2 %, making some reported results in published papers over-estimated.

Table 2.1 Coding and pooling strategies. The functions f and g are explicited below

	x_1		x_j		x_N
b_1	$u_{1,1}$	⋯	$u_{1,j}$	⋯	$u_{1,N}$
	⋮		⋮		⋮
b_i	$u_{i,1}$	⋯	$u_{i,j}$	⋯	$u_{i,N}$
	⋮		⋮		⋮
b_M	$u_{M,1}$	⋯	$u_{M,j}$	⋯	$u_{M,N}$

$\Rightarrow g : pooling$

$\Downarrow$
$f : coding$

and Chang [40] use a codebook of 166 regular colors defined a priori. These techniques are generally far from optimal, except in very specific applications. Usually, the codebook is learnt using an unsupervised clustering algorithm applied on local descriptors randomly selected from an image dataset, providing a set of M clusters with centers b_i. K-means is widely used in the BoW pipeline, whereas Gaussian Mixture Models are preferred with Fisher Vectors [35]. Other approaches [5, 22] try to include supervision to improve the dictionary learning. However, Coates and Ng [9] report that dictionary elements learned with "naive" unsupervised methods (k-means or even random sampling) are sufficient to reach high performances on different image datasets. What is reported in [9] is that most of the recognition performance is a function of the choice of architecture, specifically a good encoding function (i.e., sparse or soft) is required. In our experiments, we then choose to perform the codebook with a k-means algorithm. Let $\mathbf{B} = (\mathbf{b}_1, \ldots, \mathbf{b}_i, \ldots, \mathbf{b}_M)$ denote the resulting visual dictionary, where M is the number of visual codewords (clusters).

In Chatfield et al. [8], several mid-level representations including different coding and pooling methods are evaluated. In this chapter, we focus our re-implementation on one specific method: the Localized Soft Coding (LSC) approach [31]. Indeed, LSC proves to be a very competitive method, reaching very good results in Caltech-101 and Scene-15 databases.[4] Specifically, LSC is shown to be comparable or superior to sparse coding methods (e.g., [5, 48, 49]) while the encoding is significantly faster since no optimization is involved. Moreover, LSC is used with linear classifiers (see Sect. 2.3.3), making the representation adequate for dealing with large-scale problems.

Table 2.1 gives a matrix illustration of the mid-level representation extraction in the BoW pipeline, for scalar coding and pooling schemes. The set of local descriptors $\mathbf{X}$ is represented in columns, while the set of dictionary elements $\mathbf{B}$ occupies the rows.

---

[4] Note that from personal communication with the authors, we discover that the performances of 74 % in Liu et al. [31] in the Caltech-101 dataset have been obtained with a wrong evaluation metric. The level of performances that can be obtained with the setup depicted in Liu et al. [31] is about 70 % (see Sect. 2.5). However, the conclusion regarding the relative performances of LSC with respect to sparse coding remains valid.

One column of the matrix thus represents the encoding of a given local descriptor $\mathbf{x}_j$ into the codebook that we denote as $f(\mathbf{x}_j)$. In each row, aggregating the codes for a given dictionary elements $\mathbf{b}_i$ results in the pooling operation, denoted as $g(\mathbf{x}_j)$.

In LSC [31], the encoding $u_{i,j}$ of $\mathbf{x}_j$ to $\mathbf{b}_i$ is computed as follows using the k-nearest neighbors $\mathcal{N}_k(\mathbf{x}_j)$:

$$u_{i,j} = \frac{e^{-\beta \hat{d}(\mathbf{b}_i, \mathbf{x}_j)}}{\sum_{l=1}^{M} e^{-\beta \hat{d}(\mathbf{b}_l, \mathbf{x}_j)}} \tag{2.1}$$

where $\hat{d}(\mathbf{b}_i, \mathbf{x}_j)$ is the "localized" distance between $\mathbf{b}_i$ and $\mathbf{x}_j$, i.e., we encode a local descriptor $\mathbf{x}_j$ only on its k-nearest neighbors:

$$\hat{d}(\mathbf{b}_i, \mathbf{x}_j) = \begin{cases} d(\mathbf{b}_i, \mathbf{x}_j) & \text{if } \mathbf{b}_i \in \mathcal{N}_k(\mathbf{x}_j) \\ +\infty & \text{otherwise} \end{cases} \tag{2.2}$$

From the Localized Soft Coding strategy leading to $u_{i,j}$ codes, *max pooling* is used to generate the final image signature $\mathbf{Z} = \{z_i\}_{i \in \{1,\ldots,M\}}$ with:

$$z_i = \max_{j \in \{1,\ldots,N\}} u_{i,j} \tag{2.3}$$

The coding (function f) and pooling (function g) can be represented in this way:

$$f(\mathbf{x}_j, \mathbf{B}) = \mathbf{U}_j = \begin{pmatrix} u_{1,j} \\ \vdots \\ u_{i,j} \\ \vdots \\ u_{M,j} \end{pmatrix} \tag{2.4}$$

$$g(\mathbf{X}, \mathbf{B}) = \mathbf{Z} = \begin{pmatrix} z_1 \\ \vdots \\ z_i \\ \vdots \\ z_M \end{pmatrix} \tag{2.5}$$

where z_i is defined in Eq. (2.3).

In addition, spatial information is incorporated using a linear version [49] of the Spatial Pyramid Matching (SPM) Scheme [30]: signatures are computed in a multi-resolution spatial grid with three levels 1×1, 2×2 and 4×4. At the mid-level representation stage, the main parameter impacting accuracy is definitely M, the dictionary size.

2.3.3 Normalization and Learning

Once spatial pyramids are computed, we use linear SVMs to solve the supervised learning problem. The signature normalization is questionable. In Chatfield et al. [8], ℓ_2-normalization is applied, because this processing is claimed to be optimal with linear SVMs [46]. On the other hand, normalizing the data may discard relevant information for the classification task. For that reason, some authors report that ℓ_2-normalization negatively impacts performances, and therefore choose not performing any normalization, as in LSC [31] or in the sparse coding work of Boureau et al. [6]. Therefore, we propose here to experimentally evaluate the impact of the normalization policy on classification performances.

We use for all experiments the ℓ_2-regularized ℓ_1-loss linear SVM classification solver of the LibLinear library [15]. The SVM model can be written:

$$\min_{\mathbf{w},\xi,b} \|\mathbf{w}\|_2^2 + C \sum_{i=1}^{n} \xi_i$$
$$\text{subject to } y_i(\mathbf{w}^\top \mathbf{p}_i + b) \geq 1 - \xi_i, \forall i = 1, \ldots, n \quad (2.6)$$
$$\xi_i \geq 0, \forall i = 1, \ldots, n$$

where $(\mathbf{p}_i, y_i)_{i=1}^{n}$ with $y_i \in \{-1, +1\}$ are training samples, and C is a regularization parameter, which provides a way to avoid overfitting.

The regularization parameter C of the SVM can be determined on a validation set. In our experiments, we simply set it to a default value (10^5) because we did not observe improvement nor decline of accuracy for large values of C.

2.4 Pooling Fusion of Low and High Contrast Regions

Originally, local descriptors like SIFT [32] have been used to describe the visual content around keypoints. The keypoints are generally detected as high saliency image areas, where the contrast in the considered region is large, making the extraction of edge-based descriptors relevant. However, when a dense sampling strategy is used, the feature extraction becomes problematic because edge-based feature extraction is prone to noise in low contrast areas. This drawback is worsened with SIFT descriptors that are ℓ_2-normalized in order to gain robustness to illuminate variations: in the dense sampling setup, this normalization might make (noisy) descriptors be close to descriptors with very large gradient magnitude.

Two different ways to deal with low contrast regions are proposed in different publicly available libraries. A first one[5] renormalizes all the features whose norm is superior to a given threshold γ and divides the other features by γ so that all the

---

[5] Available on Svetlana Lazebnik's professional homepage: http://www.cs.illinois.edu/homes/slazebni/.

resulting features have a norm between 0 and 1. In this way, relevant patches with low contrast are preserved. Another one[6] also renormalizes all the features whose norm is superior to a given threshold γ but sets the other features to zero. From our experience, the latter method outperforms the former one in multiscale dense sampling strategies. We then choose to focus on the VLFEAT implementation and study the impact of the parameter γ on recognition performance. To the best of our knowledge, the impact of the normalization of low contrast regions on classification performance has never been explored in any prior work.

To better deal with low contrast areas in the BoW classification pipeline, we propose the following improvements: defining visual stop features (Sect. 2.4.1), and specific coding and pooling methods for low contrast regions (Sect. 2.4.2).

2.4.1 Visual Stop Feature: Thresholding Low Contrast Patches

In the context of image retrieval, Sivic and Zisserman [39] define **visual stop words** as the most frequent visual words in images that need to be removed from the feature representation. With the SIFT computation in low contrast patches, we are concerned about a specific type of problematic features that we call **visual stop features** since they arise at the feature extraction step (before the BoW computation). To overcome the problem of noisy SIFT computation, we threshold the descriptor norm magnitude. Let us consider a given SIFT feature $\mathbf{x}$ extracted in some region of an image. We apply the following post-processing to $\mathbf{x}$ so that the output of the feature computation is $\mathbf{x_p}$:

$$\begin{cases} \mathbf{x_p} = \mathbf{0} & \text{if } \|\mathbf{x}\| < \gamma \\ \mathbf{x_p} = \frac{\mathbf{x}}{\|\mathbf{x}\|} & \text{otherwise.} \end{cases} \quad (2.7)$$

As already mentioned, this post-processing for the SIFT computation is performed in some publicly available libraries, e.g., VLFEAT [44]. The idea is to set the descriptors corresponding to low contrast regions to a default value (e.g., $\mathbf{0}$), and not normalizing them in this case. This thresholding is dedicated to filter out the noisy feature computation by assigning a constant value to "roughly" homogeneous regions. The parameter γ defines the threshold up to which a region is considered homogeneous. In a given image $\mathcal{I}$, we denote as $\mathcal{X}_s$ the set of stop features: $\mathcal{X}_s = \{\mathbf{x} \in \mathcal{I} \ / \ \|\mathbf{x}\| < \gamma\}$. We also denote $\mathcal{X}_m$ the set of nonhomogeneous regions: $\mathcal{X}_m = \{\mathbf{x} \in \mathcal{I} \ / \ \|\mathbf{x}\| \geq \gamma\}$. Figures 2.3 and 2.4 illustrate some examples of visual stop features (illustrated with red circles) depending on γ, in Caltech-101 and Scene-15, respectively. We notice that patches with lowest magnitude mostly do not belong to the object to be recognized or do not belong to their discriminative parts (but may contain some relevant contextual information), supporting the relevance of the applied post-processing. We propose in Sect. 2.4.2 a specific modeling, in the BoW framework of stop features.

---

[6] Example VLFEAT [44] http://www.vlfeat.org/.

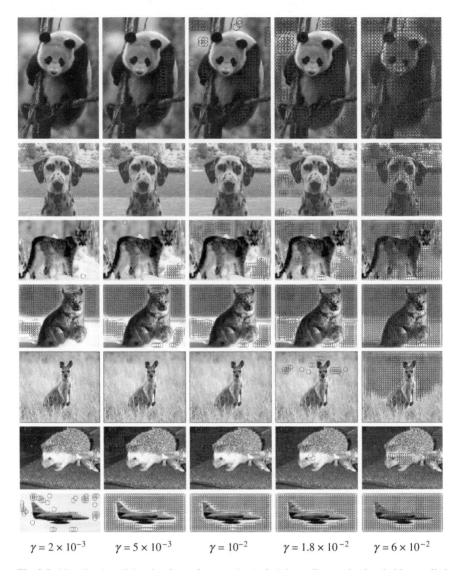

Fig. 2.3 Visualization of the visual stop features (*in circles*) depending on the threshold γ applied to the SIFT descriptor norm

2.4.2 Hybrid Image Representation

Figure 2.5 illustrates the proposed method to better deal with low contrast regions in the BoW pipeline. In particular, we adapt the coding and pooling scheme to the case of low contrast regions that are treated separately.

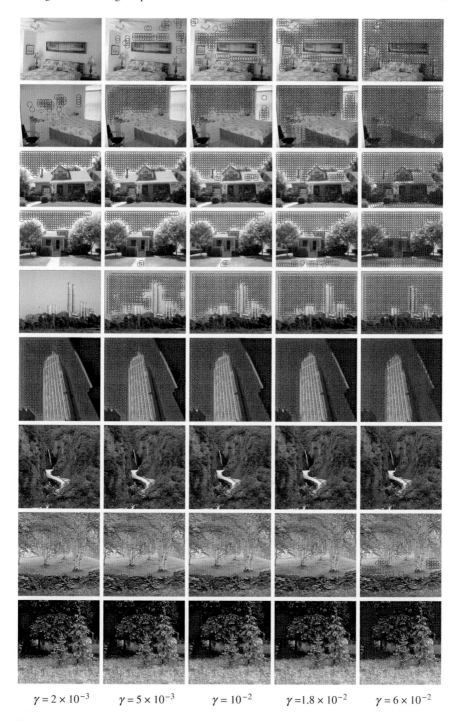

Fig. 2.4 Visualization of the visual stop features (*in circles*) depending on the threshold γ applied to the SIFT descriptor norm

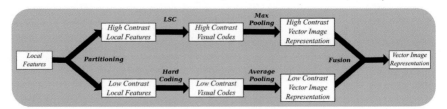

Fig. 2.5 Proposed mid-level representation in the BoW pipeline

2.4.2.1 New Dictionary Training and Feature Coding

We propose to identify a specific word in the dictionary ($\mathbf{b_0}$) to represent homogeneous regions. During codebook training, we learn the $M-1$ remaining codewords, ($\mathbf{b_1}, ..., \mathbf{b_{M-1}}$), thus excluding stop features when randomly sampling descriptors in the database. During feature encoding, we propose to hard assign each visual stop feature to the specific word corresponding to homogeneous regions ($\mathbf{b_0}$). This is sensible, since the thresholding consists of ignoring the small magnitude norm information.
For the other features, i.e., $\mathcal{X}_m$, we use the LSC method described in Sect. 2.3.2, encoding each feature on the $M-1$ "nonhomogeneous" codewords elements.

2.4.2.2 Early Fusion: Hybrid Pooling Aggregation

As described in Sect. 2.3.2, max pooling is used with LSC because it achieves better classification performances than average pooling. For visual stop features, however, since hard assignment is performed, the corresponding pooled value z_0 for the word representing homogeneous regions $\mathbf{b_0}$ using max pooling would be binary. Thus, it would only account for the presence/absence of homogeneous regions in the image. Using average pooling instead seems more appropriate: the pooled value then incorporates a statistic estimation of the ratio of low contrast regions in the image that is much more informative than the binary presence/absence value. We thus follow a hybrid pooling strategy, using average pooling for $\mathcal{X}_s$ and max pooling for $\mathcal{X}_m$. Both representations are then concatenated into a global descriptor before normalization and learning. This early fusion scheme is applied in each bin of the SPM pyramid independently.

Our hybrid pooling BoW pipeline has the following advantages: (1) The codebook can be learned only for features of $\mathcal{X}_m$, resulting in a richer representation of $\mathcal{F}_m$ for the same number of training samples; (2) The hard assignment to $\mathbf{b_0}$ for $\mathcal{X}_s$ is relevant since each homogeneous region should not be encoded in the "nonhomogeneous" codewords; (3) The encoding of $\mathcal{X}_s$ is substantially faster than using the standard LSC method, since the automatic assignment avoids the (approximate) nearest neighbor search that dominates the computational time; (4) The average pooling strategy applied to the homogeneous codeword $\mathbf{b_0}$ incorporates a richer information about the ratio of homogeneous regions in the image. This feature that must vary among different classes, can therefore be capitalized on when training the classifier.

Figure 2.6 illustrates the computation of our BoW representation for a given input image.

2.5 Experiments

Before evaluating our hybrid method, we first report an exhaustive quality assessment of the BoW strategy.

2.5.1 Datasets and Experimental Setup

Experiments are proposed on two widely used datasets: Caltech-101 [18] and Scene-15 [30]. Caltech-101 is a dataset of 9,144 images containing 101 object classes and a background class. Scene-15 contains 4,485 images of 15 scene categories.

A fixed number of images per category (30 for Caltech-101 and 100 for Scene-15) is selected to train models and all the remaining images are used for test. The reported accuracy is measured as the average classification accuracy across all classes over 100 splits. For each class, the accuracy is measured as the percentage of images of the class that are correctly assigned to the class by the learned classifier. All the images are resized to have a maximum between width and height set to 300 pixels.

Like Chatfield et al. [8], we only extract SIFT descriptors. We use a spatial stride of between 3 and 8 pixels (corresponding to the sampling density), and at 4 scales for the multiscale, defined by setting the width of the SIFT spatial bins to 4, 6, 8 and 10 pixels respectively. The default spatial stride is 3 pixels. When referring to monoscale, we set the width of the spatial bins to 4 pixels, with a default spatial stride of 8 pixels. SIFT descriptors are computed with the vl_phow command included in the VLFEAT toolbox [44], version 0.9.14, for the following experiments (Sect. 2.5.2). Apart from the stride and scale parameters, the default options are used. In Sect. 2.5.4, monoscale patches are extracted with the default vl_dsift command designed for monoscale extraction.

For LSC implementation, Liu et al. [31] use $\beta = 1/(2\sigma^2) = 10$ (Eq. 2.1) with normalized features. Since the norms of VLFEAT features are equal to 512 (instead of 1 as the descriptors used in [31]), we set $\sigma \simeq 115$ and the number of nearest neighbors $k = 10$ (Eq. 2.1) to be consistent with [31].

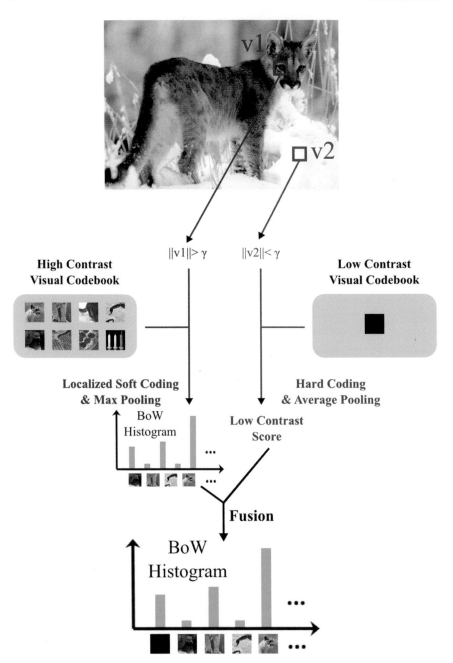

Fig. 2.6 Pooling fusion applied on an image: local descriptors are separated in two different sets depending on their norm. High contrast and low contrast regions are processed in two different coding and pooling schemes. Their resulting BoWs are concatenated in a single BoW

Table 2.2 Classification results on Caltech-101 dataset with 30 training images per class

Spatial stride	Scaling	Codebook size	Accuracy (no norm)	Accuracy (ℓ_2-norm)
8	Monoscale	800	70.07 ± 0.96	70.46 ± 1.04
6	Monoscale	800	71.64 ± 0.99	72.01 ± 0.96
3	Monoscale	800	72.45 ± 1.05	72.73 ± 0.99
8	Monoscale	1700	71.67 ± 0.93	71.95 ± 0.90
8	Monoscale	3300	72.13 ± 0.99	72.50 ± 0.97
8	Multiscale	800	73.35 ± 0.89	73.83 ± 0.96
8	Multiscale	1700	75.34 ± 0.92	75.97 ± 0.86
8	Multiscale	3300	76.91 ± 0.98	77.02 ± 0.94
3	Multiscale	800	73.81 ± 0.95	73.99 ± 0.86
3	Multiscale	1700	75.72 ± 1.13	76.00 ± 0.94
3	Multiscale	3300	77.23 ± 1.02	77.47 ± 0.99
3	Multiscale	6500	78.00 ± 1.05	78.46 ± 0.95

2.5.2 Bag-of-Words Pipeline Evaluation

We study in Table 2.2 the results of the BoW pipeline using the LSC coding method for Caltech-101 dataset. The main parameters studied are the codebook size, the spatial stride, the mono/multiscale strategy, and the normalization.

We selected the most important combinations between all the possibilities. First, one can notice that multiscale is always above monoscale results. In monoscale setup, we do not investigate too many combinations. The best results are 72.73 % for a small spatial stride with normalization. The codebook size of 3,300 also gives good results. Compared to the classical performance of 64 % of the BoW SPM [30], it is remarkable to see how a careful parametrization including normalization of a BoW soft pipeline may boost the performances up to 9 %.

These trends are fully confirmed in the multiscale setting. The best score of 78.46 % is obtained with a small spatial stride of 3, multiscale, and a dictionary of size 6,500 with ℓ_2-normalization. The soft BoW pipeline outperforms the advanced methods presented in Chatfield et al. [8], the Fisher Kernel method (reported at 77.78 %), and the LLC (reported at 76.95 %) with the same multiscale setup and a codebook of 8,000 words (for LLC). It is also above the score of Boureau [6], where the best result reported using sparse coding is 77.3 %. They use a very high dimensional image representation and a costly sparse coding optimization, with a monoscale scheme but a two-step aggregating SIFT features.

Table 2.3 reports the experimental results on Scene-15. They are all consistent with the experiments on Caltech-101. The best result of 83.44 % is also obtained for a multiscale scheme, a small spatial stride of 3, and a large dictionary of size 6,800 with normalization. This score is still slightly better than the Boureau one of 83.3 % [6], but remains below state-of-the-art results for that database.

These experiments confirm that the parameters mentioned in Sect. 2.3 may significantly improve the recognition. A small spatial stride with multiscale, a large

Table 2.3 Classification results on Scene-15 dataset with 100 training images per class

Spatial stride	Scaling	Codebook size	Accuracy (no norm)	Accuracy (ℓ_2-norm)
8	Monoscale	1000	78.72 ± 0.62	78.96 ± 0.60
6	Mnoscale	1000	79.53 ± 0.65	79.74 ± 0.65
3	Monoscale	1000	79.74 ± 0.61	80.05 ± 0.67
8	Monoscale	1700	79.98 ± 0.61	80.29 ± 0.58
8	Monoscale	3400	80.61 ± 0.61	81.16 ± 0.57
8	Multiscale	1000	79.59 ± 0.63	80.12 ± 0.56
8	Multiscale	1700	80.91 ± 0.56	81.25 ± 0.54
8	Multiscale	3400	82.01 ± 0.72	82.39 ± 0.60
3	Multiscale	1000	79.74 ± 0.60	80.14 ± 0.59
3	Multiscale	1700	81.03 ± 0.65	81.23 ± 0.60
3	Multiscale	3400	82.17 ± 0.73	82.42 ± 0.59
3	Multiscale	6800	82.66 ± 0.62	83.44 ± 0.55

codebook and a proper normalization of the spatial pyramid is the winning cocktail for the BoW pipeline. However, the accuracy improvement is more impressive for Caltech-101 (reaching very high performances) than for Scene-15.

2.5.3 Distribution of Gradient Magnitudes

A distribution in Caltech-101 and Scene-15 of the gradient magnitudes of patches in a monoscale setup is illustrated in Fig. 2.7. In Caltech-101, about 6% of patches have a gradient magnitude smaller than 10^{-4} and 40% of patches have a feature norm greater than 0.05. In Scene-15, less than 1% of patches have a magnitude close to 0. This difference compared to Caltech-101 comes from the fact that almost no fully homogeneous region exists in Scene-15 (whereas some images in Caltech-101 contain uniform background).

We study in Sect. 2.5.4 the impact of the parameter γ (in Eq. 2.7) on classification performance with our proposed strategy.

2.5.4 Evaluation of Our Strategy

We evaluate here the classification performances of our early fusion detailed in Sect. 2.4. First, we study the impact of γ (Eq. 2.7). Figure 2.8a shows the evolution of the classification performances depending on γ on Caltech-101 database, in both monoscale and multiscale settings. The results are largely impacted when γ varies: the performances can be improved up to 3% for the monoscale setup using $\gamma \simeq 10^{-2}$ compared to the default value. The same trend appears for the multiscale setting.

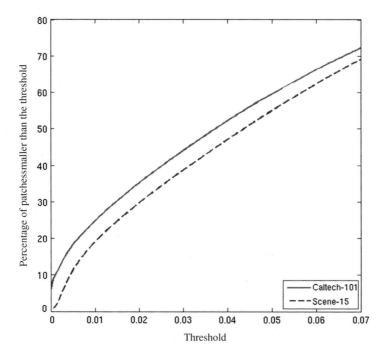

Fig. 2.7 Percentage of patches with gradient magnitude smaller than a given threshold in a monoscale setup

For Scene-15 dataset (Fig. 2.8b), the conclusion differs: in a multiscale setting the performances can be slightly improved, whereas the best result is obtained for $\gamma = 0$ with monoscale features. This may be explained by the fact that in object recognition (particularly on Caltech-101), the patches with lowest magnitude usually do not describe the object to be recognized and belong to the background (see Fig. 2.3).

Second, we evaluate the specific encoding and pooling method for low contrast regions described in Sect. 2.4.2. We provide two gradual evaluations (see Fig. 2.9). The proposed changes improve performances in Caltech-101 database, in both monoscale (Fig. 2.9a) and multiscale settings (Fig. 2.9b). For the multiscale setup, the performances are in addition more robust to γ variations. For the monoscale setup, the average pooling outperforms the max pooling method, validating the idea that enriching the homogeneous regions pooling with a nonbinary value can favorably impact performances. This is not the case in the multiscale experiments, probably because fewer homogeneous regions are extracted in such a setup (due to the increase of the region size), making the statistical estimate of the homogeneous regions ratio less reliable.

Finally, if we use the best setting of parameters with a codebook of 10^4 words, we obtain the score of **79.07 ± 0.83**% on Caltech-101 dataset and **83.83 ± 0.59**% on Scene-15 with our fusion scheme over low/high contrast regions. The reported

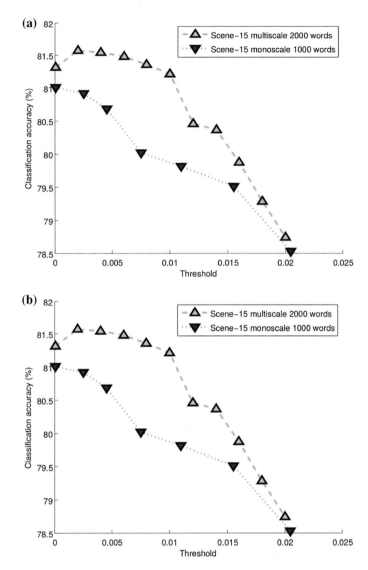

Fig. 2.8 Accuracy of the normalized LSC model as the threshold under which features are set to 0 varies **a** on Caltech-101, **b** on Scene-15

result on the Caltech-101 benchmark is comparable to the best published score [22] following the standard BoW pipeline (scalar coding and pooling + SPM), for a single descriptor type and linear classification. Note that the results obtained by Duchenne et al. [13] or Feng et al. [19] are obtained with methodological tools (resp. graph matching and pooling learning) that are complementary to our method. Therefore, a combination is expected to further boost performances.

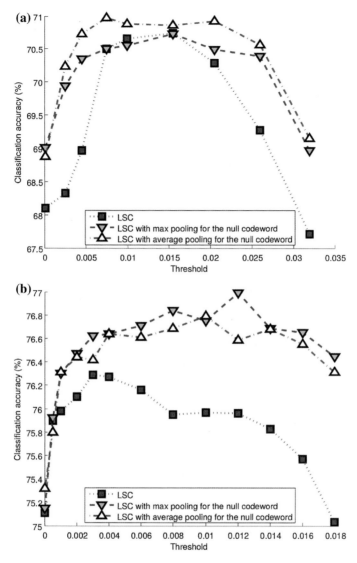

Fig. 2.9 Accuracy of the normalized LSC strategies on Caltech-101 **a** monoscale setup with a codebook of 1000 words, **b** multiscale setup with a codebook of 2000 words

2.6 Conclusions

In this chapter, we have studied in detail the different components of the BoW model in the context of image classification. Particularly, we have shown that several low (sampling rate, multiscale) and mid-level (codebook size, normalization) parameters have an impact on recognition. The codebook size and mono/multiscaling are definitely the most significant parameters as they allow to describe images with richer or denser information. The sampling rate is more significant in monoscale setup as it allows to increase the number of descriptors to represent images; this number is small in monoscale setup.

We have also investigated some early fusion methods that process low and high contrast regions separately. We have proposed a novel scheme to efficiently embed low contrast information into the BoW pipeline. This scheme is more robust than classic methods to the choice of threshold under which SIFT descriptor are normalized. This results from the fact that meaningful high contrast regions are not mixed with noisy low contrast regions. Finally, our strategy obtains state-of-the-art performances on Caltech-101 and very good results on Scene-15 dataset.

References

1. Avila S, Thome N, Cord M, Valle E, de Araujo A (2011) Bossa: extended bow formalism for image classification. In: Proceedings of the IEEE international conference on image processing (ICIP)
2. Bach FR, Lanckriet GR, Jordan MI (2004) Multiple kernel learning, conic duality, and the SMO algorithm. In: Proceedings of the twenty-first international conference on machine learning (ICML)
3. Bay H, Ess A, Tuytelaars T, van Gool L (2008) SURF: speeded Up robust features. Comput Vis Image Underst (CVIU) 110(3):346–359
4. Benois-Pineau J, Bugeau A, Karaman S, Mégret R (2012) Spatial and multi-resolution context in visual indexing. In: Visual Indexing and Retrieval, pp 41–63
5. Boureau Y-L, Bach, F, LeCun Y, Ponce J (2010) Learning mid-level features for recognition. In: Proceedings of the IEEE conference on computer vision and pattern recognition (CVPR)
6. Boureau Y-L, Le Roux N, Bach F, Ponce J, LeCun Y (2011) Ask the locals: multi-way local pooling for image recognition. In: Proceedings of the IEEE international conference on computer vision (ICCV)
7. Boureau Y-L, Ponce J, LeCun Y (2010) A theoretical analysis of feature pooling in vision algorithms. In: Proceedings of the international conference on machine learning (ICML)
8. Chatfield K, Lempitsky V, Vedaldi A, Zisserman A (2011) The devil is in the details: an evaluation of recent feature encoding methods. In: Proceedings of the British machine vision conference (BMVC)
9. Coates A, Ng A (2011) The importance of encoding versus training with sparse coding and vector quantization. In: Proceedings of the 28th international conference on machine learning (ICML)
10. Cord M, Cunningham P (2008) Machine learning techniques for multimedia: case studies on organization and retrieval. Machine learning techniques for multimedia, cognitive technologies. Springer, Heidelberg
11. Cortes C, Vapnik V (1995) Support-vector networks. Mach Learn 20(3):273–297
12. Dalal N, Triggs B (2005) Histograms of oriented gradients for human detection. In: Proceedings of the IEEE conference on computer vision and pattern recognition (CVPR)
13. Duchenne O, Joulin A, Ponce J (2011) A graph-matching kernel for object categorization. In: Proceedings of the IEEE international conference on computer vision (ICCV)
14. Everingham M, Zisserman A, Williams C, Van Gool L (2007) The PASCAL visual obiect classes challenge 2007 (VOC2007) results. Technical Report, Pascal Challenge
15. Fan RE, Chang KW, Hsieh CJ, Wang XR, Lin CJ (2008) LIBLINEAR: a library for large linear classification. J Mach Learn Res (JMLR) 9:1871–1874

16. de Avila Fontes SE, Thome N, Cord M, Valle E, de Albuquerque Arajo A (2013) Pooling in image representation: The visual codeword point of view. Comp Vis Image Underst 117(5):453–465
17. Fei-fei L (2005) A bayesian hierarchical model for learning natural scene categories. In: Proceedings of the IEEE conference on computer vision and pattern recognition (CVPR)
18. Fei-Fei L, Fergus R, Perona P (2004) Learning generative visual models from few training examples: an incremental bayesian approach tested on 101 object categories. In: Proceedings of the IEEE conference on computer vision and pattern recognition (CVPR) workshop on GMBV
19. Feng J, Ni B, Tian Q, Yan S (2011) Geometric ℓ_p-norm feature pooling for image classification. In: Proceedings of the IEEE conference on computer vision and pattern recognition (CVPR)
20. Gehler P, Nowozin S (2009) On feature combination for multiclass object classification. In: Proceedings of the IEEE international conference on computer vision (ICCV)
21. van Gemert J, Veenman C, Smeulders A, Geusebroek JM (2010) Visual word ambiguity. IEEE Trans Pattern Anal Mach Intell (TPAMI) 32(7):1271–1283
22. Goh H, Thome N, Cord M, Lim J-H (2012) Unsupervised and supervised visual codes with restricted Boltzmann machines. In: Proceedings of the European conference on computer vision (ECCV)
23. González-Díaz I, Buso V, Benois-Pineau J, Bourmaud G, Megret R (2013) Modeling instrumental activities of daily livinf in egocentric vision as sequences of active objects and context for Alzheimer disease research. In: ACM multimedia workshop on multimedia information indexing and retrieval for healthcare
24. Grauman K, Darrell T (2005) The pyramid match kernel: discriminative classification with sets of image features. In: Proceedings of the IEEE international conference on computer vision (ICCV)
25. Harris S, Stephens M (1988) A combined corner and edge detector. In: Proceedings of the 4th Alvey vision conference, pp 147–151
26. Jégou H, Douze M, Schmid C, Pérez P (2010) Aggregating local descriptors into a compact image representation. In: Proceedings of the IEEE conference on computer vision and pattern recognition (CVPR)
27. Karaman S, Benois-Pineau J, Mgret R, Bugeau A (2012) Multi-layer local graph words for object recognition. In: Proceedings of the international conference on multimedia modeling
28. Kavukcuoglu K, Sermanet P, Boureau Y-L, Gregor K, Mathieu M, LeCun Y (2010) Learning convolutional feature hierachies for visual recognition. In: Proceedings of advances in neural information processing systems (NIPS), pp 1090–1098
29. Krizhevsky A, Sutskever I, Hinton G (2012) Imagenet classification with deep convolutional neural networks. In: Proceedings of advances in neural information processing systems (NIPS), pp. 1106–1114
30. Lazebnik S, Schmid C, Ponce J (2006) Beyond bags of features: spatial pyramid matching for recognizing natural scene categories. In: Proceedings of the IEEE conference on computer vision and pattern recognition (CVPR)
31. Liu L, Wang L, Liu X (2011) In defense of soft-assignment coding. In: Proceedings of the IEEE international conference on computer vision (ICCV)
32. Lowe D (2004) Distinctive image features from scale-invariant keypoints. Int J Comput Vis (IJCV) 60:91–110
33. Mikolajczyk K, Schmid C (2004) Scale and affine invariant interest point detectors. Int J Comput Vis (IJCV) 60(1):63–86
34. Mironica I, Uijlings J, Rostamzadeh N, Ionescu B, Sebe N (2013) Time matters! capturing variation in time in video using fisher kernels. In: Proceedings of the 21st ACM international conference on multimedia
35. Perronnin F, Dance CR (2007) Fisher kernels on visual vocabularies for image categorization. In: Proceedings of the IEEE Conference on computer vision and pattern recognition (CVPR)
36. Perronnin F, Sánchez J, Mensink T (2010) Improving the fisher kernel for large-scale image classification. In: Proceedings of the European conference on computer vision (ECCV)

37. Serre T, Wolf L, Bileschi S, Riesenhuber M, Poggio T (2007) Robust object recognition with cortex-like mechanisms. IEEE Trans Pattern Anal Mach Intell (TPAMI) 29:411–426
38. Sharma G, Jurie F, Schmid C (2012) Discriminative spatial saliency for image classification. In: Proceedings of the IEEE conference on computer vision and pattern recognition (CVPR)
39. Sivic J, Zisserman A (2003) Video google: a text retrieval approach to object matching in videos. In: Proceedings of the IEEE international conference on computer vision (ICCV)
40. Smith JR, Chang S-F (1997) VisualSEEk: a fully automated content-based image query system. In: Proceedings of the fourth ACM international conference on Multimedia, ACM, pp 87–98
41. Snoek C, Worring M, Hauptmann A (2006) Learning rich semantics from news video archives by style analysis. ACM Transa Multimedia Comput Commun Appl (TOMCCAP) 2(2):91–108
42. Thériault C, Thome N, Cord M (2013) Extended coding and pooling in the HMAX model. IEEE Trans Image Process 22(2):764–777
43. van de Sande KEA, Gevers T, Snoek CGM (2010) Evaluating color descriptors for object and scene recognition. IEEE Trans Pattern Anal Mach Intell (TPAMI) 32(9):1582–1596
44. Vedaldi A, Fulkerson B (2008) VLFeat: an open and portable library of computer vision algorithms. http://www.vlfeat.org/
45. Vedaldi A, Gulshan V, Varma M, Zisserman A (2009) Multiple kernels for object detection. In: Proceedings of the IEEE international conference on computer vision (ICCV)
46. Vedaldi A, Zisserman A (2011) Efficient additive kernels via explicit feature maps. IEEE Trans Pattern Anal Mach Intell (TPAMI) 34:480–492
47. Vig E, Dorr, M, Cox DD (2012) Space-variant descriptor sampling for action recognition based on saliency and eye movements. In: Proceedings of the European conference on computer vision (ECCV)
48. Wang J, Yang J, Yu K, Lv F, Huang T, Gong Y (2010) Locality-constrained linear coding for image classification. In: Proceedings of the IEEE conference on computer vision and pattern recognition (CVPR)
49. Yang J, Yu K, Gong Y, Huang T (2009) Linear spatial pyramid matching using sparse coding for image classification. In: Proceedings of the IEEE conference on computer vision and pattern recognition (CVPR)
50. Zhang H, Berg AC, Maire M, Malik J (2006) SVM-KNN: discriminative nearest neighbor classification for visual category recognition. In: Proceedings of the IEEE conference on computer vision and pattern recognition (CVPR)
51. Zhou X, Yu K, Zhang T, Huang TS (2010) Image classification using super-vector coding of local image descriptors. In: Proceedings of the european conference on computer vision (ECCV)

Chapter 3
Hierarchical Late Fusion for Concept Detection in Videos

Sabin Tiberius Strat, Alexandre Benoit, Patrick Lambert, Hervé Bredin and Georges Quénot

Abstract Current research shows that the detection of semantic concepts (e.g., animal, bus, person, dancing, etc.) in multimedia documents such as videos, requires the use of several types of complementary descriptors in order to achieve good results. In this work, we explore strategies for combining dozens of complementary content descriptors (or "experts") in an efficient way, through the use of late fusion approaches, for concept detection in multimedia documents. We explore two fusion approaches that share a common structure: both start with a clustering of experts stage, continue with an intra-cluster fusion and finish with an inter-cluster fusion, and we also experiment with other state-of-the-art methods. The first fusion approach relies on a priori knowledge about the internals of each expert to group the set of available experts by similarity. The second approach automatically obtains measures on the similarity of experts from their output to group the experts using agglomerative clustering, and then combines the results of this fusion with those from other

S. T. Strat (✉) · A. Benoit · P. Lambert
LISTIC—University of Savoie, Annecy, France
e-mail: Sabin-Tiberius.Strat@univ-savoie.fr

A. Benoit
e-mail: Alexandre.Benoit@univ-savoie.fr

P. Lambert
e-mail: Patrick.Lambert@univ-savoie.fr

S. T. Strat
LAPI—University "POLITEHNICA" of Bucharest, Bucharest, Romania

H. Bredin
CNRS-LIMSI, Orsay, France
e-mail: Herve.Bredin@limsi.fr

G. Quénot
UJF-Grenoble 1 / UPMF-Grenoble 2 / Grenoble INP / CNRS, LIG UMR 5217,
38041Grenoble, France
e-mail: Georges.Quenot@imag.fr

methods. In the end, we show that an additional performance boost can be obtained by also considering the context of multimedia elements.

3.1 Introduction

During the last few years, society has witnessed a great increase in the amount of multimedia information, in the form of image, audio and video documents. This has created a demand for solutions aimed at automatically analyzing and organizing this content, in order to give the users the possibility to retrieve particular multimedia elements by browsing and searching the database. Formulating searches in humanly understandable *concepts* requires that the database be indexed according to such terms, which creates the need for *automatic semantic indexing* tools.

Many advances have taken place in recent years on the topic of concept detection in multimedia collections with the goal of semantic indexing and there are several well-known, publicly available datasets on which researchers can test and compare their different algorithms. For example, the Pascal VOC (visual object categories) challenge focuses on detecting objects in static images [12], the MediaEval series of benchmarks is dedicated to evaluating algorithms for multimedia access and retrieval in videos accompanied by metadata, therefore focusing even on human and social aspects of multimedia tasks [19], while the TRECVid[1] series of workshops proposes several video-only analysis tasks, such as semantic indexing and surveillance event detection [24].

A basic framework for semantic indexing on a multimedia dataset consists of extracting content descriptors from the samples (e.g., images or video shots), then training supervised classifiers on each of these descriptors. This produces, for each available descriptor and for each associated classification method, a set of classification scores that describe the "likeliness" of each sample to contain a given target concept. When possible, such scores can be calibrated as probabilities for the samples to contain the target concept.

We call an *expert* any method able to produce a set of likeliness scores for multimedia samples to contain a given target concept. Such scores can then be used to produce a ranked list of the samples the most likely to contain this concept. A combination of a content descriptor and a supervised classification method constitute an *elementary expert*. These steps are represented by the "Descriptor computation and optimization" and "Supervised classification" blocks in Fig. 3.1 (this figure illustrates the entire processing chain that we use in our experiments, which will be explained in more detail later on).

As several content descriptors and several supervised classification methods can be considered, many elementary experts can be built. So far, information coming from different elementary experts is not jointly exploited, as experts are treated independently. However, different types of elementary experts, each based on different

---

[1] TREC Video Retrieval Evaluation, http://trecvid.nist.gov/.

3 Hierarchical Late Fusion for Concept Detection in Videos

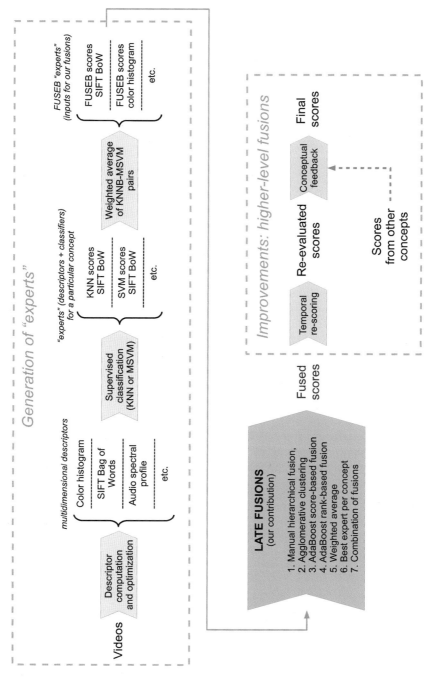

Fig. 3.1 The semantic indexing processing chain used by the IRIM group [4], in which our contribution (late fusion approaches) is integrated

aspects of the multimedia samples (such as colors, textures, contour orientations, motion or sounds, etc.), give *complementary* information.

Several aspects of complementarity can be discussed. The first is *inter-concept complementarity*, which means that a certain expert (based on a certain type of content descriptor) can give very good results for a particular semantic concept, yet perform poorly for another concept. For example, on the TRECVid SIN video dataset, the concept "*Football*" is better detected by experts using trajectory descriptors than by those using SIFT Bag-of-Words descriptors, or vice-versa. The concept "*Bridges*" is better detected with SIFT Bag-of-Words than with trajectories. There is no single expert which is systematically the best for all target concepts.

The second aspect of complementarity is *intra-concept complementarity*, which means that even if two (or more) experts have modest performances for a particular concept, their combination can produce a *higher level expert* that often performs better than any of its input elementary experts. This is especially true when one of the elementary experts detects the concept better in some situations (corresponding to some of the multimedia samples where the concept is present), while the other expert works better in the rest of the situations (the rest of the samples where the concept is present), which means that there is *complementarity at the context level*.

Because of these observations, for the sake of universality and in order to exploit complementary information, many systems rely on the combination of a large set of experts (up to 100+), each based on different descriptors or descriptor versions, and using various supervised classification algorithms.

The work described here focuses on the next step in the semantic indexing pipeline, immediately following the (multiple) supervised classification: the combination by *late fusion* of a large battery of complementary experts. The goal is to exploit their complementarity as well as possible for boosting the concept detection performance as far as possible.

The rest of the chapter is structured as follows: Section 3.2 reviews the relevant state of the art; Sect. 3.3 explains the motivation of the presented work; Sect. 3.4 describes the proposed approaches; Sect. 3.5 describes some additional improvements to the proposed approaches; Sect. 3.6 presents the experiments carried out and the obtained results; and Sect. 3.7 draws some conclusions and gives some perspectives.

3.2 State of the Art

Semantic concept detection in multimedia elements starts with computing descriptors. In the case of video datasets, we can have many types of descriptors, such as Bags-of-Words of local features (SIFT [20], SURF [5] or other type), color histograms, trajectories [2] or audio descriptors, with more examples given in Sect. 3.6.2. On such a descriptor, for a particular target concept, a supervised classification algorithm is trained and applied (such as K-nearest neighbors, support vector machines

(SVM) with various kernels, artificial neural networks, gaussian mixture models, etc.), obtaining an elementary expert [4].

Most often, combining information from several experts improves the correct recognition rates of semantic concepts. Experts can be combined at several stages within the processing chain: *Early fusions* combine descriptors before the classification step, while *late fusions* combine the outputs of supervised classifiers.

Early fusions can be as simple as concatenating two or more multidimensional descriptors, but for better results, the fact that descriptor dimensions may have values in different ranges, that descriptors may have varying numbers of dimensions and that descriptors may have varying importances for a certain concept needs to be taken into account. In [48], early fusion is performed by computing the distance between two videos as a weighted average of distances between different descriptors. In [44], a multichannel approach is used to combine a trajectory descriptor (movements from one frame to the next) and trajectory-aligned descriptors (histograms of oriented gradients, histograms of optical flow, motion boundary histograms) as input for a SVM with a χ^2 kernel, by measuring the distance between videos as the average of distances between channels (input descriptors).

Late fusions can be as simple as averaging the output scores from classifiers based on different descriptors (averaging different experts), or can be more complex, taking into account the inter-dependencies of scores from different experts like it is done with Choquet's integral [10]. An additional level of supervised classification can also be trained on the set of experts, however this can lead to over-fitting which degrades results, and averaging output scores generally gives results just as good (or better) with less computational cost. In [48], late fusion is done by averaging output scores from different experts, but in their approach, early fusion performed better than late fusion . They also experimented with a combination of early and late fusion (double fusion) which was shown to generally outperform both the early and late fusion . In general, late fusions perform best when the experts being fused are complementary, as it was shown by [23].

In [50], a visual classifier and two textual classifiers are combined using methods from belief theory, in the context of image classification. Classifier output probabilities are first converted into consonant mass functions, and then these mass functions are combined in the belief theory using Dempster's rule [36] or the Average rule. Both rules gave significantly better results than classifiers taken independently, with Dempster's rule performing better for challenging classes.

There can also be intermediates between early fusions and late fusions. With regard to SVM classifiers, Multiple Kernel Learning (MKL) can be considered a sort of intermediate fusion. Instead of using a single kernel function for the SVM, several kernels can be combined (either working on the same data or on different data) to improve classification results [14]. For example, the multichannel approach in [44] can be regarded as a MKL problem.

In [27], an early fusion, an intermediate fusion and three late fusions are used to combine static, dynamic and audio features for activity recognition using hierarchical hidden Markov models. The early fusion is a concatenation of descriptors, while the late fusions combine confidence scores from separate classifiers. The intermediate fusion, which gives the best results in their context, considers each modality as a stream of measurements and each state of the HMM models separately the observations of each stream by a Gaussian mixture, each stream being weighted depending on the activity in question.

Fusion strategies for detecting a concept can also concern themselves with how to deal with data imbalance problems (such as in TRECVid Semantic Indexing task, where most of the concepts have many more negative-labeled examples than positive ones) or which features or descriptors are more relevant for that concept. In [48], a Sequential Boosting SVM inspired from bagging and boosting approaches is used. Bagging [7] means splitting the training database into several subparts (when there are many more training negatives than positives, the positives may be kept common to all subparts) and training a classifier on each subpart; at recognition, the outputs from those classifiers are combined (averaged) to improve the result. Boosting strategies such as AdaBoost [13, 34] train a strong classifier by combining (through weighted average) results from many weak classifiers. In TRECVid, late fusions based on AdaBoost have been used in [8, 43, 45].

In the context of the TRECVid Semantic Indexing (SIN) task and as part of our participation with the IRIM group, we opt for the use of late fusion approaches (in a concept-per-concept manner), because an early fusion would mean training supervised classifiers on very high-dimensional descriptors, which is not trivial. Late fusions are easier to apply, because they fuse simple classification scores, not complex multidimensional descriptors, and in the case of TRECVid SIN, it was shown in [4] that late fusions also give better results. As inputs for the late fusion, we have a battery of (50+) *experts*, which are classification scores for each of the multidimensional descriptors (and their versions), on each video shot and each concept. A similar fusion context is described in [9], where experts are generated from a large number of video descriptors on which different classification algorithms are applied, the classifier that yields the best result for each descriptor is retained and the resulting experts are combined in a late fusion approach.

3.3 Choice of Late Fusion Strategy

When looking for an effective combination of experts, several questions arise. Should we use them all in the fusion process, or just the best ones? Does combining two experts always yield better results than the two of them taken separately? Should we weigh them differently in case one is much better than the other? Tackling a similar problem, Ng and Kantor [23] proposed a method to predict the effectiveness of their fusion approach and concluded:

> *Schemes with dissimilar outputs but comparable performance are more likely to give rise to effective naive data fusion.*

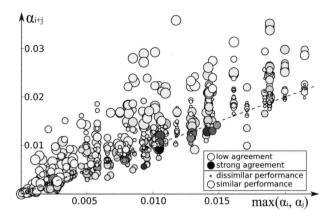

Fig. 3.2 Average precision gains when combining experts that have various performances and various agreement rates. Each *circle* represents an expert pair. The *x*-axis corresponds to max (α_i, α_j), the average precision of the best expert from a pair. The *y*-axis indicates α_{i+j}, the average precision of the combination of a pair. *Dark* (resp. bright) *grey circles* indicate that experts i and j strongly agree (resp. disagree) in their rankings. The circle diameter is directly proportional to the ratio of the average precisions in the pair α_i/α_j (where $\alpha_i < \alpha_j$)

where the *similarity* between two experts *outputs* can be measured as the Spearman rank correlation coefficient [17]—and *naive data fusion* should be understood as fusion by sum of normalized scores.

3.3.1 Fusion of Two Experts

In order to validate the conclusion of Ng and Kantor [23] in the case of concept detection in videos, we drove a simple experiment whose outcome is summarized in Fig. 3.2.

Given a set of $K = 50$ experts trained for the detection of a given concept, and an estimation of their performance (average precision) α_k on the *TRECVid 2010 Semantic Indexing* task [24], we considered all pairs (i, j) of experts and evaluated the performance of their fusion by weighted sum of normalized scores:

$$\mathbf{x} = \alpha_i \cdot \mathbf{x}_i + \alpha_j \cdot \mathbf{x}_j \tag{3.1}$$

As most circles are above the $x = y$ line (i.e., $\alpha_{i+j} > \max(\alpha_i, \alpha_j)$), Fig. 3.2 clearly shows that the weighted sum fusion from Eq. (3.1) is the most beneficial for experts that tend to disagree on their rankings but have similar average precisions (bright, large circles). This means that the gain is maximum when we have *intra-concept complementarity, at the context level*, as discussed in Sect. 3.1.

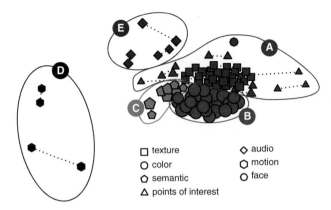

Fig. 3.3 Similarity of experts trained for the detection of concept *Computers*. Each *node* represents an expert, and *edges* represent the similarity between them (we only display some of the edges). The *dotted edges* represent experts which derive from the same descriptors, but use different classifiers

3.3.2 Communities of Experts

We have given an example for two experts, however, as described in Sect. 3.6.2, the final objective is to combine a large collection of (50+) experts. The difference between those experts mostly comes from the type of descriptors they rely on, and partly from the type of classifiers trained on top of these descriptors.

We expect experts relying on similar descriptors to generate similar outputs and therefore strongly agree with each other. We ran an additional set of preliminary experiments in order to verify this hypothesis—as illustrated in Fig. 3.3.

In Fig. 3.3, each expert is represented by a node and similar experts (according to their Spearman rank correlation coefficient [17]) are positioned closer to each other using a standard spring-layout algorithm. It appears that some kind of community structure naturally emerges, with several groups of experts being more strongly connected internally than with the outside of their group.

This is partly due to the type of descriptors used internally by the experts (denoted by the shape of the nodes). For instance, experts based on color descriptors (circles) seem to agglutinate, as do experts based on audio descriptors (diamonds). Finally, the size of a node is directly proportional to the performance (average precision) of the corresponding expert. Therefore, best performing experts (i.e., larger nodes) also tend to agglutinate as they provide rankings that are closer to the reality—therefore closer to each other.

We also used the so-called *Louvain* algorithm to automatically detect communities of experts in this graph [6, 22]. With no objective groundtruth to compare with, it is difficult to evaluate the detected communities. However, looking at Fig. 3.3 and the five detected communities (A to E), it seems that the *Louvain* algorithm did a goodjob at finding communities related to the type of descriptors on which experts

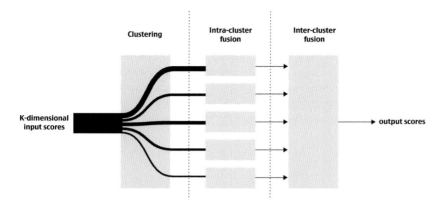

Fig. 3.4 Basic principle of our main fusion approaches: K input experts are available, which are clustered based on similarity into several groups, followed by an intra-cluster fusion and an inter-cluster fusion. Figure from [42]

are based. In particular, a dotted edge between a pair of experts indicates that they are based on the very same descriptors and they only differ in the classifier they rely on. None of these pairs is split into two different communities.

3.3.3 Hierarchical Fusion of Multiple Experts

Based on the effects noted in Sect. 3.3.1, and as illustrated in Fig. 3.4, the late fusion approaches that we propose share the following general framework:

- First, experts are grouped based on similarity into clusters of similar experts. This grouping can either be done manually, using external knowledge about the internal workings of each expert (e.g., grouping all experts that use color descriptors), or automatically, as it was done in Sect. 3.3.2.
- Then, intra-cluster fusions are performed, in which the experts from each cluster are fused. This balances the quantity of experts of each type, avoiding the case when numerous similar experts dominate the others (because some groups may be very numerous, while other groups may only have a few or even a single expert), and also helps to reduce classification "noise" within the group.
- Last, an inter-cluster fusion is performed, in which the different clusters (which are complementary because they contain experts of different types) are fused together. This gives the main performance boost due to complementarity, based on the remark of Ng and Kantor [23] and on our preliminary tests from Sect. 3.3.1.

3.4 Proposed Approaches

Our goal is to combine information coming from different experts in a way close to the optimum, so that the gain from complementarity is maximized. Following their successful use in our previous work [42], we propose two approaches: one that relies on manually grouping experts, and the other that determines the group and the weight of each expert automatically. Our main fusion approaches are the following:

- *Manual hierarchical fusion*: Expert groups are chosen manually, in a hierarchical manner, based on how the expert was obtained. There are several fusion levels, corresponding to the levels of the expert hierarchy.
- *Agglomerative clustering*: This is our automatic approach; experts are fused progressively based on similarity into groups, followed by inter-group fusion. We also extend this approach compared to what was done in [42].

3.4.1 Manual Hierarchical Fusion

The manual hierarchy was designed according to a high-level knowledge about the descriptors and the classifiers. The main principle considered is to fuse first descriptors or classifiers that are expected to be closer considering their nature or principle of operation. The manual hierarchy incorporates more levels than the automatic ones, with branches with different depths. In practice, we fused first the output of all the available machine learning algorithms for each descriptor (e.g., kNN and SVM, corresponding to block "*Weighted average of KNNB-MSVM pairs*" in Fig. 3.1). We then fuse different variants of the same descriptor (e.g., BoW of the same local descriptor but with different dictionary sizes). Afterwards, we fuse the experts corresponding to different image spatial decompositions (pyramid) if available. Finally, the last level concerns descriptors of different types within the same modality (e.g., color, texture, interest points, percepts, or faces) and descriptors from different modalities (audio and visual).

Various experiments with manually defined hierarchies suggested that going from the most similar to the most different was a good strategy. These experiments also showed that the best results are obtained when using as many combinations as possible of descriptors and machine learning algorithms. Even combinations with low performance can contribute to a global performance increase, especially if they are complementary to better ones.

Late fusion was performed at all levels using a weighted arithmetic mean of normalized scores. Several other and more complex methods were tried but produced no or very small improvements. Three weighting strategies were considered: uniform (simple arithmetic mean), MAP based (simple function of the Mean Average Precision of the different inputs), and direct optimization by cross-validation. Cross-validation experiments showed that in the early stages, uniform weighting was preferable for robustness while in latter stages MAP-based or directly optimized weighting provided better results.

3.4.2 Agglomerative Clustering and Extensions

The original version of this approach from [42] is based on grouping and fusing experts progressively based on similarity, until a minimum similarity threshold is reached; it clusters experts into groups and performs intra-group fusion at the same time. Because of this functioning, we call this fusion method *agglomerative clustering*. After this step, inter-group fusion is performed to obtain the fused result.

Compared to what was done in [42], we extend this agglomerative clustering approach by also performing, in parallel, four additional fusions: two versions of AdaBoost fusions inspired from [8, 43, 45], one weighted arithmetic mean of experts, and the best expert for each concept. At the end, the results of the five fusions are combined by choosing, for each semantic concept, the fusion method among the five that gave the best result for that concept on the training set.

We will first present the original approach, utilizing only agglomerative clustering, and then we will detail the other fusions with which we compare and also extend the agglomerative clustering.

3.4.2.1 Agglomerative Clustering of Experts

The agglomerative clustering fusion method treats each semantic concept independently, and for each concept, applies the following steps:

1. *Relevance of experts estimation*: The relevance of each of the input elementary experts is estimated on the training set, for the concept in question. The relevance is measured as the average precision of the expert normalized with respect to chance (the result of randomly choosing samples). An expert with a relevance of one means that it performs just as poorly as chance.
2. *Selection of experts*: Experts with a relevance less than one are thrown away, because they are irrelevant to the concept in question. Experts with a relevance eight times smaller than that of the best are also thrown away, in order not to "pollute" the best expert with others that are much worse. This second selection is not critical, neither is its threshold, but using it tends to reduce performance degradation from fusion for the (very few) concepts that have an extremely good best expert.
3. *Iterative fusion*: Some of the retained experts are highly correlated, so we look for the pair of experts *with the maximum correlation* and fuse it into a single expert (through arithmetic mean). The correlation between the resulting expert and the remaining ones is updated, and the process is repeated. The iterative fusion stops when a sufficiently correlated pair of experts can no longer be found. The iterative fusion corresponds to the first two steps in Fig. 3.4, as it groups and fuses similar experts at the same time (progressively, as pairs of highly correlated experts are found).
4. *Weighted arithmetic mean*: The iterative fusion does not give a large gain, because it only groups and fuses *similar* experts. The main performance boost comes now,

when we fuse *different* groups via a weighted mean of experts. The weights are given by the average precisions (for the current concept on the training dataset) of the experts from the previous step. A single expert is obtained, the result of our agglomerative clustering fusion approach. This weighted arithmetic mean corresponds to the last step in Fig. 3.4.

The correlation measure used in the iterative fusion step is the Pearson product-moment correlation coefficient ρ of the raw classification scores. $\rho \in [-1; 1]$, with values in the range of 0.6–1 corresponding to high correlation. In order to fuse a pair of experts, not only does the correlation coefficient for the classification scores of *all* samples need to be at least 0.75 (the two experts give similar information on a global scale), but also the correlation coefficient for the scores of *only the positive* samples must be at least 0.65 (to ensure that the two experts tend to detect more or less the same true positives of the semantic concept being analyzed). The constraint related to positives was added again with regards to the remark of Ng and Kantor, as at this stage, we want to group similar (not very complementary) experts; also, without this constraint, because of the imbalance between positives and negatives, the scores for negatives would have dominated the correlation measure.

The goal of iterative fusion is to balance the contribution of each family of experts, as we will see in Sect. 3.6.2 that some families are very numerous, while other families are small. This method is automatic and avoids needing to specify the families manually, making it practical for often-changing expert sets and for automatically grouping experts of similar types but from different contributors. The groups formed by the iterative fusion correspond in a large degree to the expectations based on descriptor type.

In addition to the agglomerative clustering fusion, we also experiment with other fusion approaches and with combining the results from these different fusion approaches, as described in the following.

3.4.2.2 AdaBoost Score-Based Fusion

AdaBoost [13], short for "adaptive boosting", is an algorithm that constructs a strong expert through a weighted average of a large number of weak experts. AdaBoost functions properly when each of the weak experts is at least slightly better than chance, and when the different involved experts are complementary (they each correctly classify different parts of the dataset). This is very much the case of TRECVid, where we have a large battery of experts, most of them not having spectacular individual performance (but better than chance), organized into complementary families.

The AdaBoost algorithm that we use is inspired from the original one in [13] with adaptations for TRECVid. It is very similar to that of [45], however they applied it in a different context of TRECVid. It is also very similar to that used by Tang and Yanai [43] in the 2008 edition of TRECVid, but they did not use it on such a large battery of experts as we do in our experiments.

For a particular concept, given the training set $(x_1, y_1), \ldots, (x_m, y_m)$ where x_i are the multimedia samples, and $y_i \in \{0, 1\}$ is the groundtruth of the sample x_i (0 if it does not contain the concept, 1 if it does), the algorithm that we use is the following:

1. We initialize a set of weights D_1 where $D_1(i)$ is the weight of sample x_i:

$$D_1(i) = \begin{cases} \frac{0.5}{nPos}, & \text{if } y_i = 1 \text{ (a positive sample)} \\ \frac{0.5}{nNeg}, & \text{if } y_i = 0 \text{ (a negative sample)} \end{cases} \quad (3.2)$$

where $nPos$ and $nNeg$ are the number of positive and negative samples respectively in the training set.

2. At iteration t ($t = 1, \ldots T$), we choose the input expert h_t that minimizes the weighted classification error $\varepsilon_t = \sum_{i=1}^{m} D_t(i) I(y_i \neq h_t(x_i))$. I is called the indicator function, and it gives the cost associated to the classification result of a sample being different than the groundtruth. In our case, $I(y_i \neq h_t(x_i)) = |y_i - h_t(x_i)|$, the absolute value of the difference between the classification score (between 0 and 1) and the groundtruth (0 *or* 1).

3. Compute the weight updating factor $\alpha_t = ln\frac{1-\varepsilon_t}{\varepsilon_t}$;

4. Update the weights of the samples according to:

$$D_{t+1}(i) = D_t(i) exp(\alpha_t I(y_i \neq h_t(x_i))) \quad (3.3)$$

and normalize the weights for positive samples and for negative samples separately, so that $\sum_{i, y_i=1} = 0.5$ and $\sum_{i, y_i=0} = 0.5$ (always keep the total weight of positives and the total weight of negatives equal).

5. Repeat steps 2–4 until all input experts have been considered (each expert is only considered once).

6. At the end, the *strong expert* $H(x)$ will be a weighted sum of the weak experts chosen at each iteration t:

$$H(x) = \sum_{t=1}^{T} \alpha_t h_t(x) \quad (3.4)$$

AdaBoost works on the following principle: at each step, we select the expert that correctly classifies the multimedia samples for which the previous expert failed, this way achieving *intra-concept complementarity at the context level*. Unlike agglomerative clustering, it does not first group experts into families and then obtain complementarity between families; instead, AdaBoost tries to exploit complementarity directly by choosing, at each step, the most complementary expert.

For datasets with severe class imbalance (as is the case of the TRECVid SIN video dataset, in which, for many concepts, there are only a few tens of positives and hundreds of thousands of negatives), we have added the additional constraint that the total weight of positives and the total weight of negatives should have fixed values on 0.5 each, at every iteration, as in [45], so that the classification result for true positives would still matter in the fusion.

Also for the case of TRECVid, we performed a similar expert preselection as for the agglomerative clustering fusion: we rejected experts with relevances less than 1 or less than 8 times that of the best expert for that concept, for similar reasons as in the case of the agglomerative clustering.

3.4.2.3 AdaBoost Rank-Based Fusion

When quering a dataset for a particular concept, we receive a ranked list of multimedia samples, in descending order of their likelihood to contain the concept. Ideally, in this ranked list, all the true positives should be concentrated toward the beginning, and all the negatives should follow until the end of the list. The previous AdaBoost method was made to improve the classification scores, which would indirectly improve the ranked list. We now try to optimize directly the ranks of the true positives, by altering the indicator function (the cost function when a classification error appears).

We therefore propose the following indicator function: for a positive sample, the associated cost is equal to the number of negatives that are in front of it in the ranked list, divided by the total number of negatives; for a negative sample, the cost is zero (we don't care about its rank, as long as the positives are in front):

$$I(y_i \neq h_t(x_i)) = \begin{cases} \frac{negPreceeding}{nNeg}, & \text{if } y_i = 1 \text{ (a positive sample)} \\ 0, & \text{if } y_i = 0 \text{ (a negative sample)} \end{cases} \quad (3.5)$$

where *negPreceeding* is the number of negatives preceeding the positive sample in question in the ranked list, according to the weak expert h_t, and *n*Neg is the total number of negatives.

As with the agglomerative clustering fusion and the AdaBoost fusion based on scores, we perform similar expert selections before starting the actual fusion.

3.4.2.4 Weighted Average of Experts

As a reference for comparing the performances of the fusion methods presented so far, we consider the weighted average of the input experts, with weights given by the average precisions of experts on the training set, for the concept in question (the weights can vary from one concept to another, depending on how the experts react to the concepts). We can say that in the end, the other methods are also weighted means of experts, but with more elaborate ways of choosing the weights. We wish to compare the more elaborate methods with this simple baseline.

As with the other fusion methods presented so far, we perform similar expert selections before starting the actual fusion.

3.4.2.5 Best Expert per Concept

We add a second reference for evaluating the performance of our fusion methods, namely the best expert per concept. This method consists of simply choosing, for each semantic concept individually, the expert that gives the best average precision on the training set. This is our most basic reference when examining other methods, as the goal of fusions is to obtain gains compared to simply considering the best expert for the concept of interest.

3.4.2.6 Combining Fusions

After applying all of the previous approaches in parallel, we now dispose of a battery of five fused experts: agglomerative clustering, score-based AdaBoost, rank-based AdaBoost, weighted average and best expert per concept. Our preliminary experiments have shown that for some concepts, some (or all) of the fusion methods degrade performance on the training set when compared to simply choosing that concept's best expert. To prevent this, we propose that for each concept, we see which of the fusion methods (including the best expert per concept) performs best on the training set, and *choose* that fusion method as the final result for that concept.

3.5 Improvements: Higher-Level Fusions

After the late fusion step, we dispose, for each concept, of the classification scores on all video shots. So far, we have treated each concept independently, disregarding any relationship that may exist between concepts. Moreover, the video shots from TRECVid result from the temporal segmentation of longer videos, therefore there may also exist temporal relations between shots. The next step is to integrate this *temporal context* and *semantic context* information.

A concept that is present in a shot of a video also tends to be present in the neighboring shots of the same video due to temporal correlation. We exploit this *temporal context information* by applying the method from [30] to *temporally re-score* shots, which was shown to increase performance in this application context [30] (block "*Temporal re-scoring*" in Fig. 3.1).

After temporal re-scoring, we exploit the *semantic context information* by applying *conceptual feedback* on the classification scores with the algorithm from [16]. This exploits the semantic relations between concepts by constructing a new descriptor with 346 dimensions (exactly the number of concepts), the ith dimension of this descriptor being the classification score of the shot with the ith concept. Supervised classification is applied on this descriptor as if it were a normal descriptor, and the resulting classification scores are re-fused with the previous results (block "*Conceptual feedback*" in Fig. 3.1). This step was also shown to increase performance in our application context [16].

3.6 Experiments

3.6.1 The TRECVid Semantic Indexing Task

The work presented here has been carried out and evaluated in the context of the Semantic Indexing Task (SIN) of the TRECVid evaluation campaign. The 2013 dataset associated with this task is composed of cca. 1400 h of web video data decomposed into cca. 35,000 video documents and cca. 880,000 *shots*. Shots are short video fragments of lengths varying between a few seconds to a few tens of seconds; they generally correspond to continuous camera recordings and are expected to have a homogeneous content and they constitute natural indexing and retrieval units.

A list of 346 various concepts is also provided. These can be objects (*Bus, Tree, Car, Telephone, Chair*), actions (*Singing, Eating, Handshaking*), situations/scene types (*Waterscape, Indoor, Kitchen, Construction site*), abstract concepts (*Science/technology*), types of people (*Corporate leader, Female person, Asian people, Government leader*) or even specific people (*Hu Jintao, Donald Rumsfeld*). These concepts may or may not be present in a shot. Semantic indexing, as defined in TRECVid, consists in automatically detecting the presence of these visual concepts in video shots [37].

The dataset is split into two parts, the first one (dev or 2013d), for developing and fine-tuning semantic indexing systems, and the second one (test or 2013t) for evaluating the performances of the task participants. On the test part of the dataset, semantic indexing systems are required to produce, for each target concept, a ranked list of up to 2000 shots the most likely to contain it. The quality of the returned lists (how well the relevant shots for that concept are concentrated toward the beginning of the list) is evaluated using the *mean inferred average precision (mean infAP)* [46, 47]. Common annotations are given on the dev part for system training and assessments are provided on the test part for system evaluation.

The TRECVid SIN dataset is very challenging, for the following reasons:

- Videos come from a wide array of sources, of varying quality and content, ranging from professional news footage to amateur videos recorded with a camera phone. They can be from various environments, such as from inside a kitchen or from outside in the street or at the beach. They can be acquired in various lighting conditions, ranging from a sunny day outdoors to a dark interior of a night club.
- The large amount of concepts to be detected requires a generic approach to be used for all concepts. However, it is not easy to develop a generic system that works well enough with every concept.
- Many concepts are quite rare in the dataset; they may only appear in a few tens of shots out of the total ≈880,000, which poses a problem for training classifiers.
- For a shot to be considered as an occurrence of a concept, it is enough that the concept is present in at least one frame of the shot. However, the training annotation only says if a shot contains or does not contain a concept, but it does not say *when*

and where that concept appears. This poses a challenge because we do not know which part of the shot is relevant and needs to be described.

We have chosen to perform our experiments on this dataset because it is so challenging (e.g., the peak performances in the 2012 edition were in the order of 0.3 mean infAP [38], far from the ideal value of (1) and because, as we have participated in the task as a member of the IRIM[2] group, we have had access to a large battery of multimodal video descriptors (and corresponding experts) on which we could experiment with information fusion approaches, which is the topic of this work.

3.6.2 Elementary Experts

Recalling the processing chain from Fig 3.1, the first step for semantic indexing is to extract descriptors from the video shots. For its participation in the TRECVid challenge, the laboratories that form the IRIM group have all shared their descriptors, creating a very rich and multimodal representation of the video shots. The IRIM partners have contributed many descriptors and descriptor versions, and a full listing of them is beyond the scope of this work. Instead, we will just list some of the main descriptors, without going into details:

- A large family of color descriptors was submitted by ETIS, with color represented in the Lab color space, with an optional spatial division of the keyframe [15]. A color histogram in the RGB color space was also submitted by LIG.
- ETIS also contributed quaternionic wavelets, which are a texture descriptor, also with an optional spatial division of the keyframe [15].
- A normalized Gabor transform of the keyframe was contributed by LIG, as well as an early fusion of their RGB color histogram and this normalized Gabor transform.
- BoW descriptors based on Local Binary Patterns were contributed by LIRIS [49], and texture local edge patterns enhanced by color histograms [49] were contributed by CEALIST. Multilevel histograms of multiscale LBP with spatial pyramids were contributed by LSIS [26].
- BoW of Opponent SIFT features: contributed by LIG in versions with keypoints either from a Harris-Laplace corner detector, or from a dense grid [33]. From the same family, CEALIST contributed BoW of dense SIFT with spatial pyramids [3, 35] and LISTIC contributed BoW of dense SIFT employing retinal preprocessing [39–41].
- Vectors of locally aggregated tensors (VLAT) [21], which also deal with local SIFT features clustered on a visual vocabulary, but use a pooling mechanism different than BoW to generate image signatures, were submitted by ETIS.
- Saliency moments, a descriptor that exploits the shape and contours of salient regions [28], were submitted by EUR.

---

[2] http://mrim.imag.fr/irim/

- BoW of space-time interest points, described with histograms of oriented gradients or with histograms of optical flow, as in [18], were submitted by LIG.
- EURECOM submitted spatio-temporal edge histograms, based on temporal statistics of the (2D) MPEG-7 edge histogram.
- Descriptors based on tracking and describing faces in successive frames (face tracks) were submitted by LABRI.
- LISTIC submitted Bags of Words of trajectories for motion description.
- Audio descriptors in the form of a BoW of Mel-frequency cepstral coefficients (MFCC) were contributed by LIRIS.
- Detection scores of various semantic concepts from the ILSVC and ImageNet datasets [11] (with detectors trained on ImageNet) were submitted by XEROX [32]. From the same family of highly semantic descriptors, LIF contributed a descriptor based on detection scores for a set of 15 mid-level concepts called "percepts" [1].

Before supervised classification, most of the descriptors went through an optimization (block "Descriptor computation and *optimization*" in Fig. 3.1) consisting in applying a power transformation to normalize the values of the descriptor dimensions, followed by Principal Component Analysis (PCA) to make each descriptor more compact, and at the same time, more robust [31].

The next step was to train and apply supervised classification algorithms (classifiers) on each of the (optimized) descriptors ("*Supervised classification*" in Fig. 3.1). A classifier gives, for each concept and for each video shot, the estimated "likeliness" of the shot to contain the concept (a classification score between 0 and 1).

Two classifiers were applied to each video shot descriptor. The first one is based on a K-Nearest Neighbors search.[3] The second one, called MSVM, applies a multiple learner approach based on Support Vector Machines [29]. MSVM generally performs better than KNN, but it is more computationally expensive [4].

KNN and MSVM classifiers applied to a given descriptor constitute two different elementary experts. These can be combined (or fused) into a first level non-elementary expert. The combination can be done in a number of ways. For this first level, we use a weighted mean of classification scores, the weights between KNN and MSVM being their infAP performance estimated by cross-validation within the training (dev) set. The corresponding expert is called FUSEB; it is most often better than either KNN or MSVM. We later *use the FUSEB experts as elementary ones* for the next steps in our proposed late fusion approaches.

The most numerous family of FUSEB experts is that of ETIS color histograms in the Lab color space (12 experts), while their quaternionic wavelets family numbered nine experts. LISTIC had in total 11 SIFT-based BoW experts, some with and some without retinal preprocessing, and for five experts using trajectories. Six OpponentSIFT BoW experts from LIG were also used, as well as two more dense SIFT experts from CEALIST. There were five experts based on percepts, while the experts corresponding to the remaining descriptors from the previous list were less numerous (only one or two).

---

[3] http://mrim.imag.fr/georges.quenot/freesoft/knnlsb/index.html

Table 3.1 Mean (over all concepts) inferred average precisions of fusion approaches

	basic	+RS	+RS+CF	+RS+CF+RS
Manual hierarchical fusion	0.2576	0.2695	0.2758	0.2848
Adaboost score-based fusion	0.2500	0.2630	–	–
Adaboost rank-based fusion	0.2346	0.2534	–	–
Agglomerative clustering fusion	0.2383	0.2516	–	–
Weighted average fusion	0.2264	0.2409	–	–
Best expert per concept	0.2162	0.2367	–	–
Selected best from 5 above	0.2495	0.2631	–	–

Basic (without any post-processing), +RS (with temporal re-scoring, *temporal context* integration), +RS+CF (with RS followed by conceptual feedback, *semantic context* integration), +RS+CF+RS (+RS+CF followed by a second RS)

3.6.3 Results

All of the compared fusion methods are tested using the same input elementary experts, the FUSEB experts for the descriptors listed in Sect. 3.6.2. The classifiers are trained on 2013d and applied on 2013t. The fusions are also trained on experts from 2013d, and fusion results are evaluated on 2013t. In the case of parameter optimizations for experts or fusions, they are done in cross-validation on 2013d.

We report mean infAP averaged over a subset of 38 concepts out of the total 346, the same concepts that are used for evaluating official TRECVid SIN 2013 submissions [25].

3.6.3.1 Global Results

Table 3.1 (column "*basic*") shows the mean infAP obtained by the proposed fusion methods. The *manual hierarchical fusion* performs the best, thanks to the carefully optimized weights of experts, the additional score normalization steps between fusion stages and the manual grouping of experts that ensures more homogeneous properties within a group.

Among the automatic methods, the *Adaboost score-based fusion* performs the best, with performances not far behind the manually optimized hierarchical fusion. The *Adaboost rank-based fusion* performs less good, because the rank of a shot can vary greatly with small variations in the classification score, which makes the method more sensitive to classification noise. The *agglomerative clustering fusion* is relatively close in global results to the *Adaboost rank-based fusion*. Among the fusion methods, the *weighted average fusion* is the least good, showing that a performance boost can be obtained with more careful expert weight choosing strategies; for example, the *Adaboost score-based fusion* performs 10% better than the weighted average.

In any case, it can be seen that whatever the fusion method, the global result is always better than what would have been obtained if we would have taken, for each concept, its best expert on the training dataset (*Best expert per concept*). The *manual hierarchical fusion* is 19 % better, the *Adaboost score-based fusion* is 16 % better and the even the *weighted average* has a 5 % improvement, proving that late fusion schemes, even naive ones, generally improve concept detection performances.

The *selected best fusion* selects, for each concept, the fusion approach (among *Adaboost score-based fusion, Adaboost rank-based fusion, agglomerative clustering, weighted average* and the *best expert for that concept*) that performed the best on the training set. The *Adaboost score-based fusion* was by far chosen the most often, for 230 out of the 346 concepts, which is in agreement with it having the highest mean infAP. The *Adaboost rank-based fusion* was chosen for 60 concepts, the *agglomerative clustering* for 14 concepts and the *weighted average* for only eight concepts. For the rest of the 34 concepts, the *best expert* was chosen, because the fusions were found to degrade performances on the training dataset. Considering this, it was to be expected that the mean infAP of the *selected best fusion* would be close but slightly above that of the *Adaboost score-based fusion*. However, no global gain is observed for the emphselected best fusion, because the choices made on the training set are not always the best also for the test dataset, due to variations between the two datasets.

3.6.3.2 Concept-per-Concept Results

Moving on to a concept-per-concept analysis, Table 3.2 shows the infAP gains for the 38 semantic concepts used in the official TRECVid evaluation, when comparing the best of the automatic methods (the *Adaboost score-based fusion*) with the baseline *best expert per concept*. For the majority of concepts, the fusion gives a significant performance boost (such as for *Airplane, Bus, Hand, Running, Throwing*). For some concepts, the boost is not too high, especially for concepts that already have large infAP to start with (such as *Beach, Government leader, Instrumental musician, Skating*); this happens when the other experts do not bring any pertinent and complementary information compared to the best expert. There are only six concepts that experience performance degradations from the fusion, namely *Animal, Computers, Explosion or fire, Female face closeup, Girl and Kitchen*.

As a preliminary conclusion, we can say that fusing a large battery of complementary experts yields a significant performance increase. It is now time to examine the gains of higher-level fusions, at the temporal and semantic context levels.

3.6.3.3 Results for Higher-Level Fusions

Table 3.1, column "*RS*" shows the mean infAP after applying the temporal re-scoring algorithm described in Sect. 3.5. Our best-performing method, the *manual hierarchical fusion*, has a gain of 4.6 %, while the other methods also experience gains in the range of 5–10 %. This shows that the temporal context can also bring useful information, resulting in a performance increase for all methods.

Table 3.2 Comparison of inferred average precisions for the *best expert per concept* and the *AdaBoost score-based fusion*, for particular concepts

Concept	Best expert	AdaBoost sc.	Rel. gain (%)
Airplane	0.0573	0.0923	61
Anchorperson	0.4850	0.5988	23
Animal	0.0659	0.0078	−88
Beach	0.4658	0.4722	1
Boat or ship	0.2907	0.3083	6
Boy	0.0291	0.0316	9
Bridges	0.0372	0.0393	6
Bus	0.0273	0.0598	119
Chair	0.1621	0.2394	48
Computers	0.2647	0.1919	−28
Dancing	0.2990	0.4019	34
Explosion or fire	0.1780	0.1617	−9
Female face closeup	0.3741	0.3550	−5
Flowers	0.1752	0.1895	8
Girl	0.0462	0.0360	−22
Government leader	0.4387	0.4546	4
Hand	0.1532	0.2847	86
Instrumental musician	0.5141	0.5782	12
Kitchen	0.1072	0.0952	−11
Motorcycle	0.1778	0.2369	33
News studio	0.7213	0.8223	14
Old people	0.3719	0.4096	10
People marching	0.0388	0.0470	21
Running	0.0863	0.1405	63
Singing	0.1096	0.1459	33
Sitting down	0.0003	0.0023	667
Telephones	0.0063	0.0133	111
Throwing	0.1121	0.2506	124
Baby	0.1317	0.2234	70
Door opening	0.0369	0.0410	11
Fields	0.0753	0.1375	83
Flags	0.2607	0.2819	8
Forest	0.0911	0.1150	26
George Bush	0.6092	0.6624	9
Military airplane	0.0172	0.0381	122
Quadruped	0.0807	0.1133	40
Skating	0.4956	0.5328	8
Studio with anchorperson	0.6228	0.6871	10

After temporal re-scoring, we apply the conceptual feedback step described in Sect. 3.5 (+*RS*+*CF* in Table 3.1). Because of the significant computational cost, we limit this experiment to our best-performing method, the *manual hierarchical fusion*, for which an additional gain of 2.3% is obtained compared to the previous result. Adding a second temporal re-scoring step after the conceptual feedback

(*+RS+CF+RS*) increases results by another 3.3 %. In the end, the successive temporal re-scoring and conceptual feedback steps give an increase of 10.5 % compared to the basic approach.

3.7 Conclusion

In this work, we proposed several methods of combining dozens of input experts into better ones, and applied these methods in the context of the *TRECVid 2013 Semantic Indexing* task. We have shown that all of the methods globally outperform taking the best expert for each concept, and that more elaborate fusions can perform better than a naive weighted arithmetic mean. Two late fusion methods distinguish themselves, a manually optimized hierarchical grouping of experts and an automatic fusion based on AdaBoost, both with a relatively low computational complexity. Even though we experimented on the TRECVid SIN video dataset, these approaches are generic and can be extended to other multimedia collections as well. We have also shown that additional levels of fusions that exploit context can give an additional performance increase: in the case of a video dataset, the temporal and semantic context were tested, while for other multimedia datasets, different types of contextual fusions could be devised, for example by considering the identity of the multimedia sample's uploader, the date and time when the material was created and/or uploaded etc. In the future, we plan to extend our work to such types of multimedia datasets.

Acknowledgments This work was supported by the Quaero Program and the QCompere project, respectively funded by OSEO (French State agency for innovation) and ANR (French national research agency). The authors would also like to thank the members of the IRIM consortium for the expert scores used throughout the experiments described in this paper.

References

1. Ayache S, Quénot G, Gensel J (2007) Image and video indexing using networks of operators. J Image Video Process 2007(3):1:1–1:13. doi:10.1155/2007/56928. http://dx.doi.org/10.1155/2007/56928
2. Ballas N, Delezoide B, Prêteux F (2011) Trajectories based descriptor for dynamic events annotation. In: Proceedings of the 2011 joint ACM workshop on modeling and representing events, J-MRE '11. ACM, New York, pp 13–18. doi:10.1145/2072508.2072512. http://doi.acm.org/10.1145/2072508.2072512
3. Ballas N, Labbé B, Shabou A, Borgne L (2012) Cea list at trecvid 2012: semantic indexing and instance search. In: Proceedings of TRECVid workshop, Gaithersburg, 2012
4. Ballas N, Labbé B, Shabou A, Le Borgne H, Gosselin P, Redi M, Merialdo B, Jégou H, Delhumeau J, Vieux R, Mansencal B, Benois-Pineau J, Ayache S, Hamadi A, Safadi B, Thollard F, Derbas N, Quenot G, Bredin H, Cord M, Gao B, Zhu C, Tang Y, Dellandrea E, Bichot CE, Chen L, Benoit A, Lambert P, Strat T, Razik J, Paris S, Glotin H, Trung TN, Petrovska-Delacrétaz D, Chollet G, Stoian A, Crucianu M (2012) IRIM at TRECVid 2012: semantic indexing and instance search. In: Proceedings of the workshop on TREC video retrieval eval-

uation (TRECVid). Gaithersburg, p 12. http://hal.archives-ouvertes.fr/hal-00770258. CNRS, RENATER, several Universities, other funding bodies (see https://www.grid5000.fr)
5. Bay H, Ess A, Tuytelaars T, Van Gool L (2008) Speeded-up robust features (surf). Comput Vis Image Underst 110(3):346–359. doi:10.1016/j.cviu.2007.09.014. http://dx.doi.org/10.1016/j.cviu.2007.09.014
6. Blondel VD, Guillaume JL, Lambiotte R, Lefebvre E (2008) Fast unfolding of communities in large networks. J Stat Mech: Theory Exp 2008(10):10008. http://stacks.iop.org/1742-5468/2008/i=10/a=P10008
7. Breiman L (1996) Bagging predictors. Mach Learn 24(2):123–140
8. Cai N, Li M, Lin S, Zhang Y, Tang S (2007) Ap-based adaboost in high level feature extraction at trecvid. In: Proceedings of 2nd international conference on pervasive computing and applications, 2007. ICPCA 2007, pp 194–198. doi:10.1109/ICPCA.2007.4365438
9. Cao L, Chang SF, Codella N, Cotton C, Ellis D, Gong L, Hill M, Hua G, Kender J, Merler M, Mu Y, Smith JR, Felix XY (2012) Ibm research and columbia university trecvid-2012 multimedia event detection (med), multimedia event recounting (mer), and semantic indexing (sin) systems. In: NIST TRECVid workshop, Gaithersburg, 2012
10. Cliville V, Berrah L, Mauris G (2004) Information fusion in industrial performance: a 2-additive choquet-integral based approach. In: IEEE international conference on systems, man and cybernetics, vol 2, pp 1297–1302. doi:10.1109/ICSMC.2004.1399804
11. Deng J, Dong W, Socher R, Li LJ, Li K, Fei-Fei L (2009) ImageNet: a large-scale hierarchical image database. In: CVPR09, 2009
12. Everingham M, Gool LV, Williams CKI, Winn J, Zisserman A (2010) The pascal visual object classes (voc) challenge. Int J Comput Vis 88(2):303–38
13. Freund Y, Schapire RE (1997) A decision-theoretic generalization of on-line learning and an application to boosting. J Comput Syst Sci 55(1):119–139. doi:10.1006/jcss.1997.1504. http://www.sciencedirect.com/science/article/pii/S002200009791504X
14. Gönen M, Alpaydın E (2011) Multiple kernel learning algorithms. J Mach Learn Res 12:2211–2268. http://dl.acm.org/citation.cfm?id=1953048.2021071
15. Gosselin PH, Cord M, Philipp-Foliguet S (2008) Combining visual dictionary, kernel-based similarity and learning strategy for image category retrieval. Comput Vis Image Underst 110(3):403–417. doi:10.1016/j.cviu.2007.09.018. http://dx.doi.org/10.1016/j.cviu.2007.09.018
16. Hamadi A, Quénot G, Mulhem P (2013) Conceptual feedback for semantic multimedia indexing. In: 11th international workshop on content-based multimedia indexing (CBMI), Veszprém, 2013
17. Kendall MG (1948) Rank correlation methods. Griffin, London
18. Laptev I (2005) On space-time interest points. Int J Comput Vis 64(2–3):107–23
19. Little S, Llorente A, Rüger S (2010) An overview of evaluation campaigns in multimedia retrieval. In: Müller H, Clough P, Deselaers T, Caputo B (eds.) ImageCLEF. The information retrieval series, vol 32. Springer, Berlin, pp 507–525. doi:10.1007/978-3-642-15181-1_27. http://dx.doi.org/10.1007/978-3-642-15181-1_27
20. Lowe DG (2004) Distinctive image features from scale-invariant keypoints. Int J Comput Vis 60(2):91–110. doi:10.1023/B:VISI.0000029664.99615.94. http://dx.doi.org/10.1023/B:VISI.0000029664.99615.94
21. Negrel R, Picard D, Gosselin P (2012) Compact tensor based image representation for similarity search. In: 19th IEEE international conference on image processing (ICIP), 2012, pp 2425–2428. doi:10.1109/ICIP.2012.6467387
22. Newman MEJ (2006) Modularity and community structure in networks. Proc Nat Acad Sci U.S.A 103(23):8577–8582. doi:10.1073/pnas.0601602103. http://www.pnas.org/cgi/content/abstract/103/23/8577
23. Ng KB, Kantor PB (2000) Predicting the effectiveness of naive data fusion on the basis of system characteristics. J Am Soc Inform Sci 51:1177–1189. doi: 10.1002/1097-4571(2000)9999:9999⟨::AID-ASI1030⟩3.0.CO;2-E. http://dl.acm.org/citation.cfm?id=357868.357870

24. Over P, Awad G, Michel M, Fiscus J, Kraaij W, Smeaton AF, Quénot G (2011) Trecvid 2011—an overview of the goals, tasks, data, evaluation mechanisms and metrics. In: Proceedings of TRECVid 2011. NIST, USA, 2011
25. Over P, Awad G, Michel M, Fiscus J, Sanders G, Kraaij W, Smeaton AF, Quénot G (2013) Trecvid 2013—an overview of the goals, tasks, data, evaluation mechanisms and metrics. In: Proceedings of TRECVID 2013. NIST, USA 2013
26. Paris S, Glotin H (2010) Pyramidal multi-level features for the robot vision@icpr 2010 challenge. In: 20th International conference on pattern recognition (ICPR), pp 2949–2952. doi:10.1109/ICPR.2010.1143
27. Pinquier J, Karaman S, Letoupin L, Guyot P, Megret R, Benois-Pineau J, Gaestel Y, Dartigues JF (2012) Strategies for multiple feature fusion with hierarchical hmm: application to activity recognition from wearable audiovisual sensors. In: 21st International conference on pattern recognition (ICPR), pp 3192–3195
28. Redi M, Merialdo B (2011) Saliency moments for image categorization. In: Proceedings of the 1st ACM international conference on multimedia retrieval, ICMR '11, pp 39:1–39:8. ACM, New York. doi:10.1145/1991996.1992035. http://doi.acm.org/10.1145/1991996.1992035
29. Safadi B, Quénot G (2010) Evaluations of multi-learner approaches for concept indexing in video documents. In: Adaptivity, personalization and fusion of heterogeneous information, RIAO '10, pp 88–91. LE CENTRE DE HAUTES ETUDES INTERNATIONALES D'INFORMATIQUE DOCUMENTAIRE, Paris, 2010. http://dl.acm.org/citation.cfm?id=1937055.1937075
30. Safadi B, Quénot G (2011) Re-ranking for multimedia indexing and retrieval. In: ECIR 2011: 33rd european conference on information retrieval. Springer, Dublin, pp 708–711
31. Safadi B, Quénot G (2013) Descriptor optimization for multimedia indexing and retrieval. In: 11th International workshop on content-based multimedia indexing, CBMI 2013, Veszprem, 2013
32. Sánchez J, Perronnin F, Mensink T, Verbeek J (2013) Image classification with the fisher vector: theory and practice. Int J Comput Vis 105(3):222–245. doi:10.1007/s11263-013-0636-x. http://dx.doi.org/10.1007/s11263-013-0636-x
33. van de Sande KEA, Gevers T, Snoek CGM (2010) Evaluating color descriptors for object and scene recognition. IEEE Trans Pattern Anal Mach Intell 32(9):1582–1596. http://www.science.uva.nl/research/publications/2010/vandeSandeTPAMI2010
34. Schapire RE, Singer Y (1999) Improved boosting algorithms using confidence-rated predictions. Mach Learn 37(3):297–336. doi:10.1023/A:1007614523901. http://dx.doi.org/10.1023/A:1007614523901
35. Shabou A, Borgne HL (2012) Locality-constrained and spatially regularized coding for scene categorization. In: CVPR, pp. 3618–3625. IEEE, 2012. http://dblp.uni-trier.de/db/conf/cvpr/cvpr2012.html#ShabouL12
36. Shafer G (1976) A mathematical theory of evidence. Princeton University Press, Princeton
37. Smeaton AF, Over P, Kraaij W (2009) High-level feature detection from video in TRECVid: a 5-year retrospective of achievements. In: Divakaran A (ed) Multimedia content analysis. Theory and applications. Springer, Berlin, pp 151–174
38. Snoek CGM, van de Sande KEA, Habibian A, Kordumova S, Li Z, Mazloom M, Pintea SL, Tao R, Koelma DC, Smeulders AWM (2012) The mediamill trecvid 2012 semantic video search engine. In: Proceedings of the TRECVid workshop. http://www.science.uva.nl/research/publications/2012/SnoekPTRECVid2012a
39. Strat S, Benoit A, Lambert P (2013) Retina enhanced sift descriptors for video indexing. In: 11th International workshop on content-based multimedia indexing (CBMI), pp. 201–206. doi:10.1109/CBMI.2013.6576582
40. Strat S, Benoit A, Lambert P, Caplier A (2012) Retina-enhanced surf descriptors for semantic concept detection in videos. In: 3rd International conference on image processing theory, tools and applications (IPTA), 2012, pp 319–324. doi:10.1109/IPTA.2012.6469557
41. Strat ST, Benoit A, Lambert P, Caplier A (2013) Retina enhanced surf descriptors for spatio-temporal concept detection. In: Multimedia tools and applications, pp 1–27. doi:10.1007/s11042-012-1280-0. http://dx.doi.org/10.1007/s11042-012-1280-0

42. Strat T, Benoit A, Bredin H, Quenot G, Lambert P (2012) Hierarchical late fusion for concept detection in videos. In: Andrea Fusiello VMRC (ed.) Proceedings of computer vision—ECCV 2012. workshops and demonstrations, Part III, Lecture notes in computer science (LNCS), vol 7585. Springer, Berlin, pp 335–344. doi:10.1007/978-3-642-33885-4_34. http://hal.archives-ouvertes.fr/hal-00732740. Oral session 1: WS21—Workshop on information fusion in computer vision for concept recognition OSEO (French State agency for innovation) and ANR (French national research agency)
43. Tang Z, Yanai K (2008) UEC at TRECVID 2008 high level feature task. In: In: Proceedings of the workshop on TREC video retrieval evaluation (TRECVID). Gaithersburg. http://www-nlpir.nist.gov/projects/tvpubs/tv8.papers/uec.pdf NULL
44. Wang H, Kläser A, Schmid C, Cheng-Lin L (2011) Action recognition by dense trajectories. In: IEEE conference on computer vision and pattern recognition. Colorado Springs, pp 3169–3176. http://hal.inria.fr/inria-00583818
45. Wu L, Guo Y, Qiu X, Feng Z, Rong J, Jin W, Zhou D, Wang R, Jin M (2003) Fudan university at trecvid 2003. In: Notebook of TRECVid
46. Yilmaz E, Aslam JA (2006) Estimating average precision with incomplete and imperfect judgments. In: Proceedings of the 15th ACM international conference on Information and knowledge management, CIKM '06, pp 102–111. ACM, New York. doi:10.1145/1183614.1183633. http://doi.acm.org/10.1145/1183614.1183633
47. Yilmaz E, Kanoulas E, Aslam JA (2008) A simple and efficient sampling method for estimating AP and NDCG. In: Proceedings of the 31st annual international ACM SIGIR conference on research and development in information retrieval, SIGIR '08. ACM, New York, pp 603–610. DOI http://doi.acm.org/10.1145/1390334.1390437. http://doi.acm.org/10.1145/1390334.1390437
48. Zhang L, Jiang L, Bao L, Takahashi S, Li YAH (2011) Informedia@trecvid 2011: Surveillance event detection. In: TRECVid video retrieval evaluation workshop, Gaitherburg
49. Zhu C, Bichot CE, Chen L (2013) Image region description using orthogonal combination of local binary patterns enhanced with color information. Pattern Recogn. 46(7):1949–1963. doi:10.1016/j.patcog.2013.01.003. http://dx.doi.org/10.1016/j.patcog.2013.01.003
50. Znaidia A, Borgne HL, Hudelot C (2012) Belief theory for large-scale multi-label image classification. In: Denoeux T, Masson MH (eds.) Belief functions. Advances in soft computing, vol 164. Springer, Berlin, pp 205–212

Chapter 4
Fusion of Multiple Visual Cues for Object Recognition in Videos

Iván González-Díaz, Jenny Benois-Pineau, Vincent Buso and Hugo Boujut

Abstract In this chapter, we are interested in the open problem of meaningful object recognition in video. Recently the approaches which estimate human visual attention and incorporate it into the whole visual content understanding process have become popular. In estimation of visual attention in a complex spatio-temporal content such as video one has to fuse multiple information channels such as motion, spatial contrast, and others. In the first part of the chapter, we are interested in these questions and report on optimal strategies of bottom–up fusion in visual saliency estimation. Then the estimated visual saliency is used in pooling of local descriptors. We compare different pooling approaches and show results on rather interesting visual content: that one recorded with wearable cameras for a large-scale research on Alzheimer's disease. The results which will be shown together with conclusion demonstrate that the approaches based on the saliency fusion outperform the best state-of-the art techniques in this content.

4.1 Introduction

Object recognition or classification have sparked the interest of researchers for nearly three decades. Nowadays, this topic is one of the most active in the computer vision research community. Object recognition/classification is performed on several digital media such as pictures or videos. The recognition task is more or less obvious according to the visual scene complexity, and the object to find. It is indeed easier to find an object in a controlled environment than in a natural scene. Furthermore,

I. González-Díaz · J. Benois-Pineau (✉) · V. Buso · H. Boujut
Laboratoire Bordelais de Recherches en Informatique, LaBRI, 351, Cours de la Libération, 33405 Talence, France
e-mail: benois-p@labri.fr

I. González-Díaz
e-mail: igonzale@labri.fr

B. Ionescu et al. (eds.), *Fusion in Computer Vision*, Advances in Computer Vision and Pattern Recognition, DOI: 10.1007/978-3-319-05696-8_4,
© Springer International Publishing Switzerland 2014

in a real-life visual scene, objects can be numerous and located in the presence of cluttered backgrounds. The object recognition task is often dependent on the global semantic interpretation task. One does not seek to recognize all objects in the visual content, but only those ones which are of interest to understand the meaning of the scene or for the particular task being accomplished. The examples of such a *selective* interest are numerous: e.g., when seeking for identifying a person crossing the road, the observer will not focus on the surrounding buildings.

This book chapter addresses the recognition/classification of objects in complex visual scenes recorded using a wearable video camera. More specifically, we are willing to recognize manipulated objects of the *Instrumental Activities of Daily Living (IADL)* [1]. For such videos, the wearable camera is either set on the subject's shoulder or tied on the chest. Both camera positions give an *egocentric* point-of-view of the visual scene. This point-of-view has the advantage to be the best to catch the action happening. However, there is nobody behind the camera in charge of pointing to and therefore centering the object of interest. That is the reason why the object of interest may be located in an unexpected area of the video frame. This issue is not as usual in edited videos where objects of interest are almost always near the frame center. Fruthermore, IADL video scene is complex and might be cluttered as well. In general, although several manipulated objects are present on each frame, only one or two of them could be considered as *active*, that is, of special interest for the observer to understand the meaning of the scene. Hence, under this scenario, additional information must be integrated in the recognition framework to catch the attention of the observer and therefore to discriminate between the essential and secondary elements in the scene.

In contrast to the well-known sliding window approaches for object detection and recognition [2, 3], and due to the specific nature of the first-person view contents, we aim to drive the object recognition process using *visual saliency*. Under the particular scenario of egocentric video, there is usually a strong differentiation between active (manipulated or observed by the user wearing the camera) and passive objects (associated to background) and, therefore, spatial, temporal and geometric cues can be found in the video content that may help to identify the active elements in the scene.

Incorporation of visual saliency in video content understanding is a recent trend. The fundamental model by Itti and Koch [4] is the most frequently used. Nevertheless, other models can be proposed using priors on the content. The application of saliency modeling for object recognition on video serves for identifying areas where objects of interest are located. Then, features in these areas can be extracted for object description. Several approaches in the literature have shown the utility of human gaze tracking in the analysis of egocentric video content and, in particular, in the activity recognition task [5, 6].

This chapter proposes an object recognition system that relies on visual saliency-maps to provide more precise object representations, that are robust against background clutter and, therefore, improve the precision of the object recognition task. We further propose to incorporate the saliency maps into the well known

Bag-of-Visual-Words (BoVW) [7] paradigm for object recognition. The benefits of this approach are multiple:

- The computation of saliency maps is generic (category-independent) and therefore a common step for any object detector.
- Compared to sliding window approaches [2, 8], by looking at the salient area we can avoid much of the computational overhead due to the scanning process and therefore use more complex non-linear classifiers.
- Since the saliency maps are automatically computed in both training and test data, our method does not need bounding boxes for training, what dramatically reduces the human resources devoted to the database annotation.

We consider two differentiated scenarios of application. The first one is a *constrained* scenario in which all the subjects perform actions in the same room and, therefore, interact with the same objects: e.g. a hospital scenario in which the medical staff asks patients to perform several activities. This task can be seen as a specific object recognition problem since there is no intra-class variation between instances of a category other than this caused by the strong egomotion, changes on the viewpoint, illumination, occlusions, etc.

The second scenario, on the contrary, is *unconstrained*, and corresponds to recordings made at different locations. In this case, users interact with various instances of the same objects: e.g., in a home environment, a patient performs daily activities using his/her own utensils and devices, that probably differ from those ones available in another home. The second scenario is therefore much more difficult than the first one, due to the large intra-class variation as well as to the limited amount of training data (a few instances of each object category).

In this chapter, we will assess our method in both scenarios, showing its strength and weakness in comparison to other methods in the literature.

The remainder of the chapter is organized as follows: in Sect. 4.2 we discuss the traditional visual cues fusion for the object recognition problem. Next, in Sect. 4.3, we provide a description of the geometric-spatio-temporal cues to compute saliency maps, and present some specially tailored developments to extend their use to an object recognition task in egocentric videos. Section 4.4 introduces our saliency-based approach for object recognition. In Sect. 4.5 an in-depth evaluation is provided that assesses our model under the considered scenarios, and compares it to other state-of-the-art approaches. Finally, Sect. 4.6 draws our main conclusions and introduces our further lines of research.

4.2 Traditional Visual Cues Fusion for Object Recognition Problem

Object recognition or classification are very active research topics. Over thousands of papers have been published on these subjects during the last ten years. Doing an exhaustive state of the art is therefore unrealistic. Hence we focus on the approaches that have received the most attention and have given the most promising results. One common strategy for all these methods can be highlighted. First, the image or areas of interest is described with the most possible pertinent information. The descriptors can either be local, global, or semi-local. Next, a compact representation of the set of all the descriptors is defined. Finally, distances or similarities between these representations are computed so that the current image can be classified or compared to a database in order to obtain the recognition result. In this section, all these steps are detailed.

4.2.1 Visual Cues

In order to analyze the content of images or videos, the first step consists in extracting some features which characterize the data. This step is useful for all the applications such as Content-Based Image Retrieval (CBIR), image classification, object recognition or scene understanding. Attending to their granularity, the features can either be global, local or semi-local, and almost all of them can be easily adapted to describe particular areas of interest (salient areas) detected in the video frames. In the following we review some of the existing approaches on the topic.

4.2.1.1 Global Image Descriptors

Global image features are generally based on color cues. Indeed, color is an important part of the human visual perception. In images, the colors are encoded in color spaces (RGB, HSV, YUV etc.).

Probably the most famous global color descriptor is the color histogram. Color histograms aim at representing the distribution of colors within the image or a region of the image. Each bin of a histogram represents the frequency of a color value within this area. It usually relies on a quantization of the color values, which may differ from one color channel to another. Histograms are invariant under geometrical transformations of the region.

Color moments are another way of representing the color distribution of an image or a region of an image. The first order moment is the mean which provides the average value of the pixels of the image. The standard deviation is the second order moment representing how far color values of the distribution are spread out from each other. The third order moment, named skewness, can capture the asymmetric degree of the distribution. It will be null if the distribution is centered on the mean. Using color moments, a color distribution can be represented in a very compact way [9, 10].

Other color descriptors that can be mentioned are the Dominant Color Descriptor (DCD) introduced in the MPEG-7 standard [11] or the Color Layout Descriptor (CLD).

4.2.1.2 Local Image Descriptors

The features that have received the most attention in the recent years are the local features. The main idea is to focus on the areas containing the specially discriminative information. In particular, the descriptors are generally computed around several interest regions in the image, and are therefore often associated to an interest point detector. In the following paragraphs, we briefly introduce some local descriptors that have been broadly adopted by the computer vision community in the last few years.

Scale Invariant Feature Transform

The Scale Invariant Feature Transform (SIFT) [12] has been designed to match different images or objects of a scene. The features are invariant to image scaling and rotation, and partially invariant to change in illumination and 3D camera viewpoint. They are well localized in both the spatial and frequency domains, reducing the probability of disruption by occlusion, clutter, or noise. In addition, the features are highly distinctive, which allows a single feature to be correctly matched with high probability against a large database of features, providing a basis for object and scene recognition. There are two main steps for extracting SIFT features: the key-point localization through scale-space extrema detection and the generation of key-point descriptors. First, a scale pyramid is built by convolving the image with variable-scale Gaussians and DoG images are computed from the difference of adjacent blurred images. Interesting points for SIFT features finally correspond to local extrema of these DoG images. To determine the key-point orientation, necessary for rotation invariance, a gradient orientation histogram is computed in the neighborhood of the key-point. The contribution of each neighboring pixels is weighted by the gradient magnitude. Peaks in the histogram indicate the dominant orientations. The feature descriptor finally corresponds to a set of orientation histograms, relative to the key-point orientation, on a 4×4 pixel neighborhoods. As histograms contain eight bins, a SIFT feature is a vector of 128 dimensions. This vector is normalized to ensure invariance to illumination changes.

Speeded Up Robust Features

Although SIFT have proven to be a powerful feature in many computer vision applications, all the necessary convolutions make it computationally expensive. Hence, Speeded Up Robust Features (SURF) [13] have then been proposed as an alternative feature. This feature describes a distribution of Haar-wavelet responses within interest point neighborhood. It relies on integral images. The latter is the sum of all pixel values contained in the rectangle between the origin and the current position. SURF key-points are also extracted by scale-space analysis through the use of Hessian-matrices. Here again, the dominant orientation is extracted. It is estimated by computing the sum of Haar-wavelet responses within a sliding orientation window. In an oriented square window centered at the key-point, which is split up into 4 × 4 sub-regions, each sub-region finally yields a feature vector based on the Haar-wavelet responses, of dimension 64.

4.2.1.3 Semi-Local Image Descriptors

Most shape descriptors fall into this category. Shape description relies on the extraction of accurate contours of shapes within the image or region of interest. Image segmentation is usually fulfilled as a preprocessing stage. In order for the descriptor to be robust with regard to affine transformations of an object, quasi perfect segmentation of shapes of interest is supposed. Here, we just mention some shape descriptors but more can be found in literature. In particular, let us mention the Curvature Scale Space (CSS) descriptor [14] and the Angular Radial Transform (ART), descriptors in the MPEG-7 standard.

4.2.2 Models for Object Recognition

In this section, we will review some computational models that have been broadly adopted in computer vision tasks. We will start by presenting the BoVW which, due to its simplicity and notable performance, represents one of the most common approaches in tasks like object recognition, scene understanding or even action recognition. We will then introduce the family of sliding-window methods, that concurrently address the problems of object recognition and localization. Finally, we will present the use of saliency as an efficient alternative for the so high computational burden of the sliding window methods.

4.2.2.1 Bag-of-Visual-Words Paradigm

The descriptors presented above, and in particular SIFT and SURF, have been widely used for retrieving objects in images. Local feature extraction leads to a set of unordered feature vectors. The main difficulty of the recognition, retrieval or classification steps consists in finding a compact representation of all these features and its associated (dis-)similarity measure. An efficient approach that has been widely used is the so-called BoVW framework [15], that we now describe. The BoVW approaches have four main stages: building a visual dictionary by clustering visual features extracted from a training set of images/objects, quantifying the features, choosing an image representation using the dictionary and comparing images according to this representation. We now review these steps.

Visual Dictionary

In analogy with text retrieval, the features extracted in an image correspond to the words in a document. A visual dictionary must then be built. This is generally done by randomly selecting a sufficiently large set of features over a huge amount of images. This dictionary, $V = v_i, i = \{1, \ldots, K\}$, is then built by clustering these features into a certain number of K classes or "visual words."

Feature Quantization

The second step consists in quantizing the features extracted in an image according to the visual dictionary. Each feature from N extracted features for an image is *quantized*. This quantization is generally achieved by assigning each feature to its closest word in the dictionary V.

Pooling

Each image in the dataset can now be represented by a unique vector of K dimensions. Each dimension represents the number of times a feature appears in the image. Therefore, this vector can be seen as a histogram representing the distribution of visual words in an image. This histogram is often normalized which enables comparing images containing a different number of features. These histograms were named BoVW [15].

Instance Recognition

All images being now represented by a histogram, the last step simply consists in comparing the histograms. Obviously, when the size of the database increases this step can become very computationally expensive. The computational time also depends on the size of the dictionary which therefore needs to be chosen carefully. Several strategies have been proposed in the literature to improve the efficiency of this last step. In [15], this framework was applied with SIFT features. The vector quantization was carried out by k-means clustering, the number of clusters being chosen manually.

4.2.2.2 Sliding Window Methods

Unlike the general BoVW paradigm, sliding window methods aim to concurrently address the object recognition and localization tasks. These methods perform a window-based scanning process so that objects are intensively searched at several locations and scales in the image. Good examples of these methods can be found in the literature applied to face detection [16], pedestrian detection [8], deformable part models for general object detection [2], and even mixing BoV with the sliding window approach [3]. As mentioned, they have shown very good performance addressing the object detection and localization in images, even when an object size is very small compared to the image dimensions. However, their main drawback is the computational burden caused by the intensive scanning process, what prevents their application under many scenarios.

The Deformable Part Trained Model (DPM) [2] has been introduced in 2006 by Felzenszwalb et al. and remains today the state-of-the-art sliding window method in many applications in computer vision. It is an object detection system based on mixtures of multiscale deformable part models. At a high level, the system can be characterized by the combination of:

- Strong low-level features based on Histograms of Oriented Gradients (HOG).
- Efficient matching algorithms for deformable part-based models (pictorial structures).
- Discriminative learning with latent variables (latent SVM).

The system needs to be trained using bounding boxes on objects, allowing it to automatically compute the best probable models for the parts of an object.

The DPM has become a model of reference for object recognition and has been extended to different applications. Concerning egocentric videos, the authors in [17] proposed to extend the use of the DPM to train classifiers for activities based on the output of the well-known deformable part model [2] using temporal pyramids.

In this work, the DPM has been used for the purpose of comparing the performances with our method (see Sect. 4.5).

4.2.2.3 Saliency-Based Methods

In contrast to sliding window methods, visual saliency represents an efficient way to drive the scene analysis towards particular areas considered 'of interest' for a viewer and has become a very active trend in computer vision. The computation and use of saliency is specially appealing for the object recognition task due to its generic nature, not dependent on the particular object being detected. By using saliency, on the one hand, one can reduce the computational burden of the scanning process of sliding window methods and, on the other, filter out much of the background information from the scene analysis, thus giving more relevance to the area of interest.

The application of saliency modeling for object recognition on video serves for identifying areas where objects of interest are located. Then, features in these areas can be extracted for object description. Several approaches in the literature have shown the utility of human gaze tracking in the analysis of visual content addressing tasks such as image retrieval [18], object recognition [19], or action recognition [5, 6, 20].

This book chapter explores how an object recognition system is built in the top of visual saliency-maps that provide more precise object representations, that are robust against background clutter and, therefore, improve the precision of the object recognition task. We further propose to incorporate the saliency maps into the well known Bag-of-Visual-Words (BoVW) [7] paradigm for object recognition. As mentioned in the introduction, the use of saliency in object recognition is generic and does not depend on the particular object/concept being detected, and therefore avoids much of the computational burden of the sliding window methods by filtering our much visual information and driving the recognition process to some areas of particular interest in the scene.

4.3 Visual Saliency: A New Paradigm for Information Fusion in Object Recognition

4.3.1 Motivations

In this work, we propose to use visual saliency for detecting *active* regions of the frame. Visual saliency represents the human visual attention within a visual scene. Therefore, the saliency is well suited to distinguish active from inactive objects. Visual saliency modeling captivates researchers since the early 80s with the *Feature Integration Theory* [21] from Treisman and Gelade. This research topic is still very active. In 2012, Borji, and Itti [22] took the inventory of 48 significant saliency models. Despite the fact that the visual saliency modeling is an old research topic, object recognition frameworks using such models is a new trend [5, 23]. Most of the visual saliency models are only considering spatial information such as contrast. These models are called *spatial* and where designed at first for still pictures. There are also models called *spatio-temporal* based on the motion present in videos. Especially the Human Visual System (HVS) is highly sensitive to the relative motion. This is why applying a *spatio-temporal* saliency model in the object recognition framework is relevant to consider the temporal dimension of videos. Indeed most of the object recognition frameworks for video only process video frames separately, without taking advantage of previous and next frames. In this chapter, we also propose to improve the saliency model by adding a third saliency cue called *geometric*. Recent approaches [24] have shown that subjects tend to fixate the screen center when watching natural scene. In [25], the authors came to the same conclusion for natural edited videos.

4.3.1.1 Traditional Models for Saliency

ITTI Model

"A Model of Saliency-Based Visual Attention for Rapid Scene Analysis" from Itti et al. [4] is one of the most cited articles on topics related to the visual modeling.

The approach described in the paper is built on the biologically-plausible architecture proposed by Koch and Ullman in their "Feature Integration Theory" (FIT) [26]. The FIT tries to explain how human visual search strategies are performed. Itti's model is a bottom–up approach based on three computational stages: extraction, activation, and normalization/combination. At the extraction stage features are extracted at several spatial scales (nine in the original paper) by using dyadic Gaussian pyramids. These features are intensity, color opponents (red/green, green/red, blue/yellow and yellow/blue), and the local orientation estimated from the intensity. At the activation stage, linear "center-surround" operations are applied on the feature computed at the previous stage. These "center-surround" structures are present in the HVS to detect local spatial discontinuities, especially for detecting areas standing out from the surround. In the last stage, the saliency map is built by normalizing and combining (summing) the features processed by the "center-surround" operators. The gaze scan-path is predicted from the saliency map by using a biologically-inspired 2D "winner-take-all" neural network. Each neuron of the network receives an excitatory input from the saliency map. The first neuron to "fire" is the "winner," which means that the Focus of Attention (FOA) is shifted to the location of this neuron on the saliency map. Then area covered by the winner neuron is inhibited and the WTA process is started again. This inhibition phenomenon has been observed and measured in human visual psychophysics [27]. This approach has inspired many saliency map models such as [28–30].

Graph-Based Visual Saliency

In Graph-Based Visual Saliency (GBVS), J. Harel et al. have proposed an original method for computing visual saliency maps on still images. Their bottom–up model is based on three computational stages as many other leading approaches [31–33]:

- **Extraction** (identified as s1 in the original paper): of feature vectors from the image plane.
- **Activation** (s2): Creates an activation map (or feature map) from the feature vectors extracted at stage s1.
- **Normalization/combination** (s3): Normalization of the activation map, and combination of the maps in a single map if several activation maps have been generated at stage s2.

Most algorithms perform Stage s1 by using biologically inspired filter and stage s2 is done by multiscale feature maps subtraction. This processing simulates the action of center-surround ganglion cells located in the retina. Stage s3 is accomplished either by considering the local maximum (max-ave) as in [31] or by using iterative convolutions based on Difference-of-Gaussians (DoG) filters, or with non-linear interactions (NL) [34]. The novelty is that the author has proposed to apply graph algorithms to compute saliency maps for stages s2 and s3. Their approach defines the dissimilarity and the saliency, of features, as the edge weights of the graphs. Generated graphs are interpreted as Markov chains. The authors claim that this approach is more "organic" since it is biologically inspired with individual nodes (neurons) connected together in a structured network as in the HVS. The communication between nodes allows for the rising of emergent behavior such as the fast identification of areas that require additional processing. This method (GBVS) has been compared with the standard saliency map methods from Itti et al. [31, 35, 36], Bruce and Tsotsos [33]. The results show that the GBVS predicts the fixations better than the standard methods. The authors have also compared their approach with improved version of the standard methods considering the center bias hypothesis on photographs. Although these improved methods provided better results, the GBVS still better predicts eye fixations.

4.3.2 Our Computational Models of Visual Saliency

In order to drive the video analysis to the regions that are potentially interesting to human observers we need to model visual saliency on the basis of video signal features. In this work, we have considered three basic approaches to generate saliency maps, each of them built using a particular source of information: spatial, geometric and temporal. In the following paragraphs, we will briefly describe the method that gives place to each map.

Spatial saliency S_s: proposed in [37], it is based on various color contrast descriptors that are computed on the HSV color space, due to its closeness to human perception of color. In particular, seven local contrasts are computed, namely:

1. *Contrast of Saturation*: A contrast occurs when low and highly saturated color regions are close.
2. *Contrast of Intensity*: A contrast is visible when dark and bright colors co-exist.
3. *Contrast of Hue*: A hue angle difference on the color wheel may generate a contrast.
4. *Contrast of Opponents*: Colors located at the hue wheel opposite sides create very high contrast.
5. *Contrast of Warm and Cold Colors*: Warm colors–red, orange and yellow—are visually attractive.
6. *Dominance of Warm Colors*: Warm colors are always visually attractive even if no contrast are present in the surrounding.
7. *Dominance of Brightness and Saturation*: Highly bright and saturated regions have more chances of attracting the attention, regardless of the hue value.

The spatial saliency value $S_s(i)$ for each pixel i in a frame is computed by averaging the outputs associated to the seven color contrasts.

Temporal saliency S_t: this saliency models the attraction of attention to motion singularities in a scene. The visual attention is not grabbed by the motion itself, but by the residual motion for each pixel, e.g., the difference between the estimated motion for each pixel and the predicted camera motion based on a global parametrization.

Simply put, the process of computing a temporal saliency map is as follows: first, for each frame in the video, a dense motion map $\mathbf{v}(i)$ that contains the motion vectors in each pixel i in the image is computed using the optical flow technique described in [38].

Then, a 3×3 affine matrix A that models the global motion associated to the camera movements is computed. For that end, the well-known robust estimation method RANSAC [39] has been used in order to successfully handle the presence of outliers (e.g., areas of the image associated to objects that move differently than the camera). Furthermore, since the central area of each frame constitutes the most likely region where moving objects appear, this region is not considered for the affine matrix estimation, thus reducing the proportion of outliers.

Next, the residual motion $\mathbf{r}(i)$ is computed by compensating the camera motion:

$$\mathbf{r}(i) = \mathbf{v}(i) - A \cdot v_i \qquad (4.1)$$

where $\mathbf{x}_i$ stands for the spatial coordinates of each pixel i, $\mathbf{x}_i = (x_i, y_i, 1)^T$.

Finally, the values of the temporal saliency map $S_t(i)$ are computed by filtering the amount of residual motion in the frame. The authors of [37] reported that the human eye cannot follow objects with a velocity higher than $80°/s$ [40]. According to this psycho-visual constraints, a post-processing filter was proposed in [37] that decreased the saliency when motion was too strong. Applying this filtering stage to our first-person camera videos was however too restrictive due to the strong camera motion so that we have preferred to consider a simpler filtering stage that normalizes and computes the saliency map as follows:

$$S_t(i) = \min \left\{ \frac{\|\mathbf{r}(i)\|_2}{K}, 1 \right\} \qquad (4.2)$$

where K has been heuristically computed depending on image dimensions (H, W), as $K = \max(H, W)/10$.

Geometric saliency S_g: it follows two observations about saliency in egocentric video: on the one hand, some studies on general purpose video confirm the so-called center bias hypothesis, that is the attraction of human gaze by the geometrical center of an image [37, 41]. On the other hand, in videos recorded with wearable cameras, the camera is usually set-up to point specific areas of interest: e.g., the gaze fixation if the camera is located on glasses, or an area just in front of the human body where the hands usually manipulate objects, in case it is located on the body. Generally, central geometric saliency is dependent on the wearable camera position and might be shifted in image plane [41]. In the present research, we work on datasets with either eye-centered or body-centered camera, thus using the center-bias hypothesis. Hence, following the approach in [37], the geometric saliency map $S_g(i) = \mathcal{N}((x_0, y_0), (\sigma_x, \sigma_y))$ is computed as 2D Gaussian located at the screen center with a spread $\sigma_x = \sigma_y = 5°$.

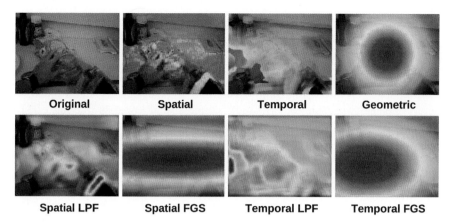

Fig. 4.1 Results of various saliency maps for one frame in GTEA dataset. The three basic techniques spatial, temporal and geometric are shown. In addition, for spatial and temporal maps, two types of postprocessing are also included (LPF and FGS)

However, this attraction may change with the camera motion. This is explained by the anticipation phenomenon [42]. Indeed, the observer of video content produced by a wearable video camera tries to anticipate the actions of the actor. The action anticipation is performed according to the actor body motion, which is expressed by the camera motion. Hence, we propose to simulate this phenomenon by moving the 2D Gaussian centered on initial *geometric saliency point* in the direction of the camera motion projected in the image plane. A rough approximation of this projection is the motion of image center computed with the global motion estimation model previously described.

Results on the basic approaches are shown in Fig. 4.1 (columns 2–4). As one can notice from the figures, spatial and temporal saliency maps show more precise localization of the objects of interest whereas the geometric approach provides a coarse approximation of the visual saliency. However, saliency information appears more scattered or disaggregated for the first two approaches, being more compact and therefore robust for the geometric technique.

For an object recognition task, we consider that the perfect saliency map is a trade-off between precision and compactness, requirement that, based on the examples, is not completely fulfilled by any of the basic approaches. Hence, to overcome this issue, we propose two extensions: (a) to incorporate a post-processing step in the spatial and temporal techniques that provides more compact saliency representations and (b) to investigate fusion schemes that successfully combine the three approaches taking advantage of their precision and compactness, respectively.

4.3.3 Postprocessing: Setting-up Suitable Saliency Maps for Object Recognition

As already mentioned, we propose to use an additional post-processing stage to obtain more compact representations for the spatial and temporal saliency. In particular, we have evaluated two methods: (a) a very simple spatial low-pass filtering using a Gaussian mask (LPF), and (b) a method that Fits a Gaussian Surface (FGS) on the original map.

The LPF approach, shown in columns 5–6 of Fig. 4.1, simply provides a smooth version of the original saliency maps. However, if the standard deviation of the spatial Gaussian is large enough, results may fulfill our requirements of compactness.

For the second approach, given the original saliency mask S, we propose to fit a Gaussian surface of the form:

$$G(x, y) = A \cdot \exp\left[-\frac{1}{2}\left(\frac{x - x_g}{\sigma_x^2} + \frac{y - y_g}{\sigma_y^2}\right)\right] \quad (4.3)$$

where $\theta = \{A, x_g, y_g, \sigma_x, \sigma_y\}$ are the parameters to be estimated in the fitting process. In practice, we minimize the square error between the two maps $e^2 = \sum_{x,y}[S(x, y) - G(x, y)]^2$ using the optimization method described in [43].

In the experimental section, we will assess the performance of both post-processing approaches.

4.3.4 Fusion Strategies for Saliency Maps

Once the basic spatial, temporal, and geometric saliency maps has been introduced, we aim to evaluate how their combination into spatio-temporal-geometric saliency masks S_{stg} might improve the representation of the area of interest in the image.

For that end, several fusion strategies have been proposed and evaluated in this work. Again, although most of them have been already proposed in [44] in a video quality assessment task, for the sake of compactness we next briefly describe their computation:

1. Multiplication (Mult): a multiplicative fusion strategy model as:

$$S_{stg}^{mult}(i) = S_s(i) \cdot S_t(i) \cdot S_g(i) \quad (4.4)$$

2. Mean: the average of the three methods as:

$$S_{stg}^{mean}(i) = \frac{1}{3}\left(S_s(i) + S_t(i) + S_g(i)\right) \quad (4.5)$$

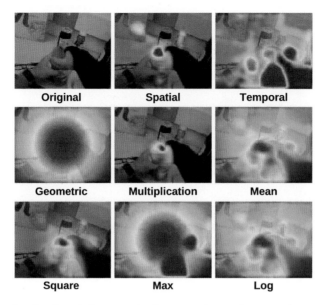

Fig. 4.2 Results of various fusion strategies for computing spatio-temporal-geometric saliency maps

3. Square: the squared Minkovsky pooling reinforced by multiplicative pooling:

$$S_{stg}^{sq}(i) = S_s(i) \cdot S_t(i) \cdot S_g(i) + \frac{1}{3}\left(S_s^2(i) + S_t^2(i) + S_g^2(i)\right) \quad (4.6)$$

4. Max: maximum pooling:

$$S_{stg}^{max}(i) = \max\left(S_s(i), S_t(i), S_g(i)\right) \quad (4.7)$$

5. Log: logarithmic combination model:

$$S_{stg}^{log}(i) = \frac{1}{3}\left(\log(1 + S_s(i)) + \log(1 + S_t(i)) + \log(1 + S_g(i))\right) \quad (4.8)$$

A visual example of the fusion strategies is shown in Fig. 4.2. In addition, all of them will be evaluated in the experimental section of this chapter.

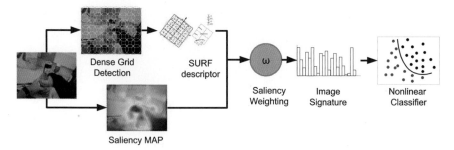

Fig. 4.3 Processing pipeline for the saliency-based object recognition in first-person camera videos

4.4 Object Recognition with Visual Saliency

4.4.1 Low-Level Feature Extraction and Description

In this section, we will describe our approach for object recognition in first-person camera videos using saliency masks. As we have already mentioned in the introduction, we aim to detect the Region of Interest (ROI) of each frame so that we can effectively build more precise image representations.

The processing pipeline of our approach is included in Fig. 4.3. We build our model on the well-known BoW paradigm [7], and propose to add saliency masks as a way to improve the spatial precision of the original Bag-of-Words approach.

For each frame in a video sequence, we extract a set of N local descriptors using a dense grid of local circular patches [45]. Based on some experiments, we have set the radius of the circular patches to 25 pixels, and the step size between each local patch of 6 pixels, thus leading to a high degree of overlapping between neighboring local regions.

Next, each local patch $n = 1, \ldots, N$ is described using a 64-dimensional SURF descriptor d_n [13], which has shown similar performance than the SIFT descriptor [12] in our experiments, whereas it is of half the dimension. Each descriptor d_n is then assigned to the most similar word $j = 1, \ldots, V$ in a visual vocabulary by following a vector-quantization process. The visual vocabulary, computed using a k-means algorithm over a large set of descriptors in the training dataset (about 1 million descriptors in our case), has a size of V visual words. As we will show in the evaluation section, we have experimented with visual vocabularies of different sizes V.

In parallel, our system generates a saliency map S of the frame with the same dimensions of the image and values in the range [0, 1] (the higher the more salient is a pixel).

4.4.2 Object Recognition with Saliency Weighting

In the traditional BoVW approach [7], the final image signature H is the statistical distribution of the image descriptors according to the codebook. This is made by first assigning each local descriptor to a visual word in the vocabulary and then computing a histogram of word occurrences by counting the times that a visual word appears in an image.

Instead of doing this hard assignment, we propose to apply what we call *saliency weighting*, a sort of soft-assignment based on saliency maps. With saliency weighting, the contribution of each image descriptor is defined by the maximum saliency value found under the circular region Ω_n associated to the index n. In other words, descriptors over salient areas will get more weight in the image signature than descriptors over non-salient areas. Therefore, the image signature is a V-dimensional vector H that can be computed as follows:

$$H_j = \sum_{n=1}^{N} \alpha_n w_{nj} \qquad (4.9)$$

where the term $w_{nj} = 1$ if the descriptor or region n is quantized to the visual word j in the vocabulary, and the weight α_n is defined as:

$$\alpha_n = \max_{s \in \Omega_n} \{S(s)\} \qquad (4.10)$$

where Ω_n represents the set of pixels contained in the nth circular region of the dense grid, and $S(s)$ is a saliency map.

Finally, the histogram H is L1-normalized in order to produce the final image signature.

It is worth stressing the difference between our weighted histogram with hard-assignments and the histogram with soft assignments previously proposed in the literature [46]. In that work, given a descriptor, a similarity measure is computed with respect to all the words in the vocabulary so that various bins of the histogram can be incremented according to these similarities. On the contrary, our method is assigning each descriptor to just one word in the vocabulary but then is weighting its contribution to the histogram using the saliency map information. In fact, if necessary, our method might be combined with the one in [46].

On the contrary, our method of saliency weighting is more similar to the spatial weighting proposed in [47] but, in our case, the weights are computed unsupervisely, without need of training data and not depending on the category to detect.

Once each image is represented by its weighted histogram of visual words, we use a non-linear classifier to detect the presence of a category in the image. In particular, we have employed a SVM classifier [48] with a χ^2 kernel, which has shown good performance in visual recognition tasks working with normalized histograms as those ones used in the BoW paradigm [49].

4.5 Experimental Results

In this section, we assess our model in various challenging datasets with egocentric videos. As we have already mentioned we aim to recognize objects under two different scenarios: constrained, in which all videos contain the same instances of the involved object categories, and the unconstrained, in which each video shows a different environment with varying instances of the object categories.

4.5.1 Datasets

We have assessed our approach with three publicly available ego-centric video datasets.

The first one is the GTEA Gaze dataset [5], which consists of 17 standard definition (640 × 480) video sequences, captured at a frame rate of 15 frames per second, and performed by 14 different subjects using Tobii eye-tracking glasses. Due to the lack of object annotations in this dataset, we have extracted and annotated 595 frames from the videos so that we can easily perform our tests over a set of still images. The whole dataset has been divided into two sets, namely: (a) the training set (294 frames) and (b) the test set (300 frames). Furthermore, we aimed to detect 15 object categories in this database. Due to its limited size, we have used this dataset to compare various system configurations.

The second dataset is the GTEA dataset [50] for Object Recognition. This dataset, recorded at 30 frames per second in 1280 × 720 definition, contains seven types of daily activities, each performed by four different subjects. In this case, the camera is mounted on a cap worn by the subject. Weak annotations are already available for this dataset. They identify active objects on each frame belonging to 16 object categories, but do not include the object location. Since all the users have been recorded in the same room interacting with the same objects, we have evaluated our constrained scenario using this dataset. For that end, we have followed the same setup described in [50], using the users 2–4 for training the algorithms and the user 1 for testing.

The third dataset used in the experiments is the ADL dataset [17], that contains videos captured by a chest-mounted GoPro camera on users performing various daily activities at their homes. The high definition videos (1280 × 960) are captured at rate of 30 frames per second and with 170° of viewing angle. In total, 27,064 frames have been accurately annotated providing bounding boxes for objects belonging to 44 categories. In our experiments, we have just considered those objects labelled as 'active' (those being interacted or observed by the users) for both training and testing purposes. This dataset is more challenging than the other two since both the environment and the object instances are completely different for each user, thus leading to an unconstrained scenario. However, we have evaluated both scenarios with this dataset: the constrained one by randomly dividing the whole set of frames into a training and test set (50–50), and the unconstrained, by doing so at the video/user level.

4.5.2 Setting-up the Final Model

In this section, we compare various system configurations. The objective is then to select the final system setup that provides the best performance, which will be compared with other state-of-the-art methods in the two envisaged scenarios.

4.5.2.1 Evaluating the Basic Approaches for Saliency Maps

We have firstly evaluated our basic approaches for generating the saliency maps. In addition, we have included two reference methods in the comparison:

1. Basic BoW (B-BoW): the BoVW approach that generates image signatures considering whole images. This method becomes the basic reference and allows us to evaluate the improvement achieved by our saliency masks.
2. BoW with Ideal Masks (I-BoW): this approach makes use of the ideal ground truth masks provided in the annotation. Since it evaluates our approach when the saliency masks correspond with the ground-truth, it constitutes the theoretical limit in its performance. It is worth noting how this ideal binary masks are used both on training and testing, thus incorporating the annotations in the whole recognition process, but omitting the aforementioned weighting scheme in the histograms computation.

The results of this study in the GTEA Gaze dataset are presented in Fig. 4.4a, that shows the Average Precision (AP) achieved by each approach at various vocabulary sizes. As one can notice from the results, for almost every technique, the performance improves until a vocabulary size of $V = 4,000$ words, after which it stabilizes. Hence, from now on, we will either remove larger vocabulary sizes from our experiments or simply consider the optimal vocabulary size of 4,000 as the final approach.

Comparing the approaches, as we expected, the I-BoW constitutes the theoretical upper bound of the method. This is logic due to the use of the ground-truth bounding boxes that, although do not correspond to the tight silhouette of the object of interest, always ensure its correct localization. Furthermore, two of the basic techniques to compute the saliency masks (geometric and temporal) already achieve slightly better results than the reference B-BoW. This is a nice consequence of the use of saliency masks, even when not specific post-processing is applied to the maps. Furthermore, the fact that the geometric saliency map is the one that achieves the best results, let us to conclude that compactness is even more important than localization precision for an object recognition task.

4.5.2.2 Techniques for Saliency Map Post-processing

In this section, we present the evaluation of the post-processing techniques described in Sect. 4.3.3. As we have already claimed, direct outputs from some saliency detectors might not be optimal for an object recognition task due to the lack of compactness.

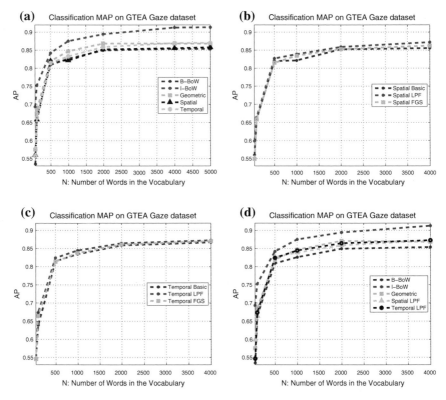

Fig. 4.4 A comparison of various configurations in the GTEA Gaze dataset and various vocabulary sizes. **a** Results of the basic saliency techniques in comparison with the two references; **b** results achieved by two post-processing techniques for the spatial saliency; **c** results achieved by two post-processing techniques for the temporal saliency; **d** a comparison between the best post-processing option (LPF) and the reference methods

Since the geometric technique already provided compact and Gaussian-shaped saliency masks, we have applied the postprocessing stage to the spatial and temporal techniques. Figure 4.4b and c respectively compare the results obtained in the GTEA Gaze dataset by the basic spatial and temporal saliency, and the two post-processing methods: Low Pass Filtering (LPF) and FGS. The improvements on the results, although not very notable, demonstrate that post-processing is important to adequate the saliency maps to the particular problem of object recognition. Furthermore, the computational cost of the LPF method, the one that achieves the best performance, is almost negligible when compared to other steps of the processing pipeline.

In addition, Fig. 4.4d shows a comparison between the LPF approach and the two reference methods. With the post-processing stage, now all the saliency methods outperform the reference B-BoW and achieve closer results to the theoretical limit I-BoW. Hence, from now on, LPF post-processing will be incorporated to every version of our approach.

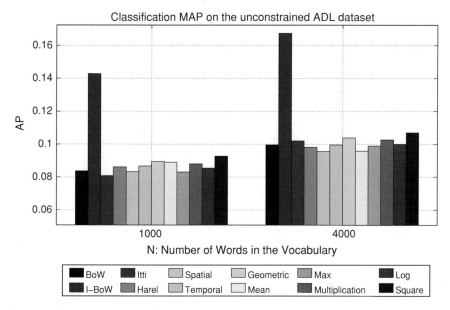

Fig. 4.5 Classification results of various strategies for fusing spatio-temporal-geometric saliency maps. Values are given at two different vocabulary sizes ($V = 1000$, $V = 4000$). Basic and reference methods are also included for comparison

4.5.3 Data Fusion Strategies for Saliency Maps

In this section, we show how spatio-temporal-geometric information can be successfully fused for object recognition. As mentioned before, rather than following the traditional approach that fuses classifiers over various features (texture, color, shape, etc.), here we experiment by fusing heterogeneous information to compute saliency maps. The resulting maps can be then easily plugged into the Bag-of-Visual-Words paradigm for object recognition.

Therefore, we have assessed several fusion approaches that have been previously described in Sect. 4.3.4, and compared them with the basic spatial, temporal and geometric saliency maps. Furthermore, we have also included the two references (B-BoW, I-BoW), as well as two well-known methods using spatial information to compute saliency maps: the fundamental model of Itti et al. [4], and the graph-based method of Harel et al. [29]. Results of this study in a subset of the ADL unconstrained dataset are shown in Fig. 4.5 for two vocabulary sizes ($V = 1000$, $V = 4000$).

The obtained results stand out the importance of the fusion strategy, which makes the difference between providing similar or even worse results than the basic approaches and yielding notably better performance. In particular, the square fusion strategy obtains particularly good performance on this dataset, outperforming both the basic saliency approaches and the rest of the fusion strategies. In particular, by using this approach we are achieving relative improvements with respect to

Table 4.1 mAP and standard deviation on ADLdataset under the constrained and unconstrained scenarios

Algorithm	Cons. mAP ± std	Uncons. mAP ± std
B-BoW	0.585 ± 0.258	0.113 ± 0.152
I-BoW	0.621 ± 0.250	0.191 ± 0.258
DPM [2]	0.341 ± 0.254	0.129 ± 0.194
Proposal	0.602 ± 0.260	0.125 ± 0.167

the reference B-BoW of a 10 and 7.43 %, for a vocabulary of size 1,000 and 4,000, respectively. Hence, we will consider this fusion strategy as the final choice for our object recognition system in ego-centric videos. The excellent results achieved in the hypothetical case in which ideal ground truth saliency maps are available (I-BoW) also help to stress the suitability of this saliency-based approach for object recognition.

Finally, it is also of interest remarking that our proposed spatio-temporal-geometric saliency map provides better performance than the two well known saliency methods (Itti and Harel). The rationale behind is that these two methods simply use spatial properties of the scene, and therefore ignore motion and geometric constraints that have turned out to be very useful to detect the area of interest of a video scene.

4.5.4 A Comparison with the State-of-the-Art

In the following, we present a comparison between our method and various approaches that have reported state-of-the-art results in the egocentric datasets. As mentioned before, we consider two scenarios of application: an easier constrained scenario and a more challenging unconstrained scenario.

4.5.4.1 Experiments Under the Constrained Scenario

As we mentioned before, the constrained scenario is that one in which all the subjects wearing cameras are recorded in the same environment and interacting with the same object instances.

Results for the ADL dataset under the constrained scenario are shown in the first column of Table 4.1 in terms of mAP (mean Average Precision), and its standard deviation (category deviation). It is worth noting that we show only the results of those objects considered as 'active' in the dataset ground-truth annotations, e.g. those objects that are either manipulated or observed by the main actor in the ego-centric video. We consider these objects as the main source of information for detecting an action, so that the rest of the visual information (background) is less relevant and only useful for horizontal tasks as context identification.

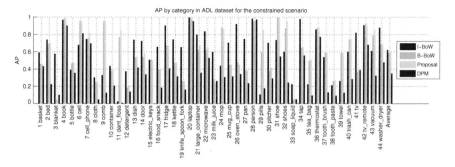

Fig. 4.6 Per-category results (AP) for the constrained scenario achieved by various methods in the ADL dataset

As we have already mentioned, to simulate the constrained environment, we have randomly divided the whole set of frames into a training and test set (50–50) without taking into account the video to which each frame belongs. In this dataset, we are comparing the performance of our approach with the reference method B-BoW, the ideal case I-BoW, and the Discriminatively Trained Part-Based Model (DPM) [2], which was the approach used by the authors of the dataset [17] to address the object recognition task.

Furthermore, in Fig. 4.6 we include detailed per-category performance. Base on these results, we can draw the following conclusions:

- Our proposal outperforms the reference B-BoW by guiding the recognition process to the salient areas of each frame. This result is consistent along almost all the categories in the dataset, and supports the idea that using visual saliency generates more accurate object representations and reduces the effect of clutter.
- The approach using ideal masks is, as expected, the one yielding the best performance. However, a deeper by category analysis shows remarkable conclusions: in general, providing an accurate localization of the object (I-BoW) helps the recognition process and improves the performance. This observation is particularly noticeable for relatively small objects such as the ones belonging to the categories 'foodsnack', 'knife_spoon_fork', 'milk_juice' or TV. However, when the objects are too small, such as the instances of 'comb,' 'dentfloss' or 'pills,' we have observed that the ground truth bounding boxes, restricted to the object and lacking any information about object context, give not enough information to successfully detect its presence. In contrast, due to the fact that the saliency maps usually cover more area in the image (object, hands, even spatial neighboring context), our proposal achieves notably better results than the I-BoW. In addition, the reference B-BoW also achieves better results than I-BoW for these classes, although its performance is still below our approach.
- The performance of the DPM is poor when compared any BoW method. From our point of view, the rationale behind is that this method has been designed to get good generalizations of object categories, what prevents from taking advantage of the

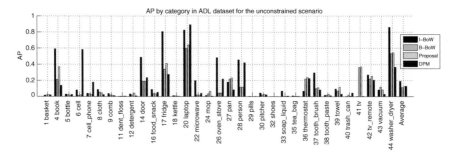

Fig. 4.7 Per-category results (AP) for the unconstrained scenario achieved by various methods in the ADL dataset. Some categories cannot be computed in this scenarios due to the lack of samples in training/test sets

high visual similarity between training and test samples in the constrained scenario. Hence, we believe that its relative performance with respect to our approach should drastically improve in the unconstrained scenario.

In addition, we have also evaluated our approach in the GTEA dataset. This dataset represents the constrained scenario in a more realistic way, due to the fact that we can take training and test samples from different videos. Hence, we have followed the same evaluation setup proposed by the authors [50]. In particular, we have developed a multiclass classifier so that each image is considered to contain just one object of interest. Our proposal achieves a global classification accuracy of 36.8 % in this dataset, which compares well with the 35 % obtained by the authors of the dataset [50] when they matched the highest detection score to the ground truth annotations.

4.5.4.2 Experiments Under the Unconstrained Scenario

The unconstrained scenario corresponds to the challenging situation in which users perform their activities at several locations, thus interacting with heterogeneous instances of the object categories. Consequently, the large intra-class variation jointly with the reduced number of object instances, are expected to lead to poor generalization in recognition process.

In our experiments, we have used the videos corresponding to half of the subjects {2, 3, 5, 7, 8, 12–14, 17, 18} for training, and the remainder videos for test.

Average results of this study are shown in the second column of Table 4.1, whereas Fig. 4.7 shows detailed per-category AP. We next draw the main conclusions of this experiment:

- As expected due to the challenging nature of this scenario, the performance is drastically lower for all the automatic approaches (from AP $\sim$ 0.6 to $\sim$ 0.10). This illustrates how challenging is the problem of object recognition when just a few instances are available for each object.

Table 4.2 Test execution times of our approach compared with the DPM implementation in [2]. We show single threading (S.T.) and multi-threading (M.T.) execution time

Algorithm	S.T.	M.T.
DPM [2]	60.4 s	10.9 s
Proposal	15.7 s	4.1 s

- Furthermore, the I-BoW, that uses ground-truth masks in test, now notably outperforms any automatic approach. This fact stresses the importance of a good previous localization of the object of interest for its localization.
- Our proposal again outperforms the basic reference (B-BoW). The improvement is once more consistent along almost all the categories.
- The DPM now achieves competitive results, even slightly superior to the ones of our proposal. As we previously stated, this technique learns object models with a high degree of generalization, which is better suited for this unconstrained rather than for the constrained scenario.

As a conclusion, we can state that our approach yields good results in the constrained scenario, outperforming state-of-the-art approaches, and obtains competitive results in the unconstrained one. In addition to the classification results, it offers two main advantages over other alternatives: (1) it does not require precise localization of objects in the training data, what minimizes the human effort in the database annotation, and (2) as we will see in the next section, its computational complexity is low when compared to sliding window methods.

4.5.5 A Study of the Computational Time

In Table 4.2, we show a comparison between the average execution times of our proposal and the DPM to run one category object-detector in a test frame. We included results using a single threading (S.T.) and multi-threading in a 2.10 GHz computer with four cores, and hyper-threading enabled.

For our proposal, the execution time comprises the generation of the saliency maps, the SURF feature extraction process, the computation of the weighted histograms, and the classification using a SVM with χ^2 kernel. It is worth noting that some of the computations for the spatial saliency map are implemented in GPU so that they cannot be translated to S.T. case (spatial saliency takes about 0.05 s per frame in the GPU). The rest of the calculus are made with the CPU under the aforementioned circumstances.

For the DPM, we run the implementation in [2], made in Matlab with optimized c routines for all the steps in the process that require most of the execution time.

As we can see in the tables, our approach shows much lower computational times in comparison with DPM. From our point of view, the rationale behind is the fact that using the saliency maps, we avoid the heavy scanning process of a sliding window approach as the DPM.

Furthermore, it is also worth noting that, since the saliency maps are automatically computed in both training and test data, our method does not need bounding boxes for training, what dramatically reduces the human resources devoted to the database annotation when compared to the DPM.

4.6 Conclusion and Perspectives

In this chapter we have presented a method for object recognition in egocentric videos. Our proposal aims to drive the recognition process using visual saliency. In particular, spatial, temporal and geometric cues found in egocentric videos are exploited to improve the object recognition, generating more precise representations of the area of interest in a frame, as well as enhancing the robustness against cluttered backgrounds.

We have also evaluated several fusion strategies to generate spatio-temporal-geometric saliency maps from their basic constituents, as well as some post-processing techniques that improve the compactness, a property that has turned out to be very important for object recognition.

In addition, rather than simply performing foreground/background segmentation to restrict the recognition process to the areas of interest, we have proposed a soft application of saliency that controls the influence of pixels in the final object representation based on their saliency. We have combined saliency with the well known BoVW paradigm by proposing a saliency weighting method to compute image signatures.

Having in mind the context of this work, which is the automatic analysis of videos for the diagnosis, assessment, maintenance and promotion of self-independence of people with dementia, we have assessed our model in two particular scenarios of interest: (a) a constrained scenario in all the subjects perform actions in the same room and, therefore, interact with the same object instances, and (b) an unconstrained scenario that corresponds to recordings made at different locations, so that users interact with various instances of the same objects.

Our experiments have shown that this method outperforms the basic BoVW model and achieves closer results to an hypothetical case in which optimal foreground masks are available in test. Furthermore, our approach compares well, and outperforms DPM and the full method in [50] under the constrained scenario. Furthermore, the computational time is less than half of the DPM one.

However, the notable decrease in performance in case of an unconstrained scenario reveals that our method needs further development. Indeed, in an unconstrained scenario the variability of object instances intra-category requires drastically new recognition approaches. Here we are in the case of "concept recognition." As we know from e.g. TRECVID challenge [51] concept recognition is a complex and open research problem and we are amongst those working on it.

Acknowledgments This research has been supported by the region of Aquitaine and the European Community's program (FP7/2007–2014) under Grant Agreement 288199 (Dem@care Project).

References

1. Pirsiavash H, Ramanan D (2012) Detecting activities of daily living in first-person camera views. In: IEEE conference on computer vision and pattern recognition (CVPR), pp 2847–2854
2. Felzenszwalb PF, Girshick RB, McAllester DA, Ramanan D (2010) Object detection with discriminatively trained part-based models. IEEE Trans Pattern Anal Mach Intell 32(9):1627–1645
3. Lampert CH, Blaschko MB, Hofmann T (2008) Beyond sliding windows: object localization by efficient subwindow search. In: IEEE computer society conference on computer vision and pattern recognition (CVPR 2008), IEEE Computer Society, Anchorage, 24–26 June 2008
4. Itti L, Koch C (2001) Computational modelling of visual attention. Nat Rev Neurosci 2(3):194–203
5. Fathi A, Li Y, Rehg JM (2012) Learning to recognize daily actions using gaze. In: Proceedings of the 12th European conference on computer vision—Volume Part I, ECCV'12, pp 314–327, Springer, Berlin, 2012
6. Ogaki K, Kitani KM, Sugano Y, Sato Y (2012) Coupling eye-motion and ego-motion features for first-person activity recognition. In: 2012 IEEE computer society conference on computer vision and pattern recognition workshops, IEEE, pp 1–7, 2012
7. Csurka G, Dance CR, Fan L, Willamowski J, Bray C (2004) Visual categorization with bags of keypoints. In: Workshop on statistical learning in computer vision, ECCV, pp 1–22
8. Dalal N, Triggs B (2005) Histograms of oriented gradients for human detection. In: Schmid C, Soatto S, Tomasi C (eds) International conference on computer vision and pattern recognition, vol 2. INRIA Rhône-Alpes, ZIRST-655, av. de l'Europe, Montbonnot-38334, pp 886–893
9. Jing F, Li M, Zhang H, Zhang B(2002) An effective region-based image retrieval framework.In: ACM international conference on multimedia, 2002
10. Long F, Zhang H, Feng D (2003) Fundamentals of content-based image retrieval. In: Multimedia information retrieval and management, 2003
11. Manjunath B, Ohm J, Vasudevan V, Yamada A (2001) Colour and texture descriptors. IEEE Trans Circ Sys Video Technol 11(6):703–715
12. Lowe DG (2004) Distinctive image features from scale-invariant keypoints. Intern J Comput Vis 60:91–110
13. Bay H, Ess A, Tuytelaars T, Van Gool L (2008) Speeded-up robust features (surf). Comput Vis Image Underst 110:346–359
14. Mokhtarian F, Suomela R (1998) Robust image corner detection through curvature scale space. IEEE Trans Pattern Anal Mach Intell 20(12):1376–1381
15. Sivic J, Zisserman A (2003) Video google: a text retrieval approach to object matching in videos. In: Proceedings of the international conference on computer vision 2:1470–1477
16. Viola P, Jones M (2001) Rapid object detection using a boosted cascade of simple features. In: 2001 IEEE computer society conference on computer vision and pattern recognition, vol 1. IEEE, Los Alamitos, pp 511–518
17. Pirsiavash H, Ramanan D (2012) Detecting activities of daily living in first-person camera views. In: 2012 IEEE conference on computer vision and pattern recognition (CVPR), IEEE, 2012
18. de Carvalho Soares R, da Silva I, Guliato D (2012) Spatial locality weighting of features using saliency map with a bag-of-visual-words approach. In: IEEE 24th international conference on tools with artificial intelligence (ICTAI), vol 1. pp 1070–1075

19. Sharma G, Jurie F, Schmid C (2012) Discriminative spatial saliency for image classification. In: IEEE conference on computer vision and pattern recognition (CVPR), pp 3506–3513
20. Vig E, Dorr M, Cox D (2012) Space-variant descriptor sampling for action recognition based on saliency and eye movements. Springer, Firenze, pp 84–97
21. Treisman AM, Gelade G (1980) A feature-integration theory of attention. Cogn Psychol 12(1):97–136
22. Borji A, Itti L (2012) State-of-the-art in visual attention modeling. IEEE Trans Pattern Anal Mach Intell 99(PrePrints), 34(9):1758–1772
23. Vig E, Dorr M, Cox D (2012) Space-variant descriptor sampling for action recognition based on saliency and eye movements. In: European conference on computer vision, 2012
24. Tatler BW (2007) The central fixation bias in scene viewing: selecting an optimal viewing position independently of motor biases and image feature distributions. J Vis 7(14):1–17
25. Dorr M, Martinetz T, Gegenfurtner KR, Barth E (2010) Variability of eye movements when viewing dynamic natural scenes. J Vis, 10(10):28
26. Koch C, Ullman S (1985) Shifts in selective visual attention: towards the underlying neural circuitry. Hum Neurobiol 4:219–227
27. Posner MI, Cohen YA (1984) Components of visual orienting. In: Bouma H, Bouwhuis DG (eds) Attention and performance X: control of language processes. Lawrence Erlbaum, Hillsdale
28. Parkhurst D, Law K, Niebur E (2002) Modeling the role of salience in the allocation of overt visual attention. Vis Res 42(1):107–123
29. Harel J, Koch C, Perona P (2007) Graph-based visual saliency. In: Advances in neural information processing systems 19. MIT Press, Cambridge, pp 545–552
30. Marat S, Ho Phuoc T, Granjon L, Guyader N, Pellerin D, Guérin-Dugué, V (2009) Modelling spatio-temporal saliency to predict gaze direction for short videos. Intern J Comput Vis 82(3):231–243
31. Itti L, Koch C, Niebur E (1998) A model of saliency-based visual attention for rapid scene analysis. IEEE Trans Pattern Anal Mach Intell 20(11):1254–1259
32. Itti L, Baldi PF (2006) Bayesian surprise attracts human attention. In: Advances in neural information processing systems, (NIPS*2005) vol 19. MIT Press, Cambridge, pp 547–554
33. Tsotsos JK, Bruce NDB (2006) Saliency based on information maximization. In: Weiss Y, Schölkopf B, Platt J (eds) Advances in Neural Information Processing Systems 18. MIT Press, Cambridge, pp 155–162
34. Itti L, Braun J, Lee DK, Koch C (1999) Attentional modulation of human pattern discrimination psychophysics reproduced by a quantitative model. In: Advances in neural information processing systems. MIT Press, Cambridge, p 1998
35. Itti L (June 2000) A saliency-based search mechanism for overt and covert shifts of visual attention. Vis Res 40(10–12):1489–1506
36. Lee DK, Itti L, Koch C, Braun J (Apr 1999) Attention activates winner-take-all competition among visual filters. Nat Neurosci 2(4):375–81
37. Brouard O, Ricordel V, Barba D (2009) Cartes de Saillance Spatio-Temporelle basées Contrastes de Couleur et Mouvement Relatif. In: Compression et representation des signaux audiovisuels, 2009
38. Farnebäck G (2000) Fast and accurate motion estimation using orientation tensors and parametric motion models. In: Proceedings of 15th international conference on pattern recognition, vol 1. IAPR, Barcelona, Sept 2000, pp 135–139
39. Fischler MA, Bolles RC (June 1981) Random sample consensus: a paradigm for model fitting with applications to image analysis and automated cartography. Commun ACM 24:381–395
40. Daly SJ (1998) Engineering observations from spatiovelocity and spatiotemporal visual models. In: IS&T/SPIE conference on human vision and electronic imagingIII:1, 1998
41. Boujut H, Benois-Pineau J, Megret R (2012) Fusion of multiple visual cues for visual saliency extraction from wearable camera settings with strong motion. In: Fusiello A, Murino V, Cucchiara R (eds) Computer vision—ECCV 2012. Workshops and Demonstrations, Lecture Notes in Computer Science, vol 7585. Springer, Berlin, pp 436–445

42. Land M, Mennie N, Rusted J (1999) The roles of vision and eye movements in the control of activities of daily living. Perception 28:1311–1328
43. Moré JJ, Sorensen DC (1983) Computing a trust region step. SIAM J Sci Stat Comput 4(3):553–572
44. Boujut H, Benois-Pineau J, Ahmed T, Hadar O, Bonnet P (2011) A metric for no-reference video quality assessment for hd tv delivery based on saliency maps. In: IEEE international conference on multimedia and expo, July 2011
45. Tuytelaars T, Lampert C, Blaschko M, Buntine W (2010) Unsupervised object discovery: a comparison. Intern J Comput Vis 88:284–302
46. Philbin J, Chum O, Isard M, Sivic J, Zisserman A (2008) Lost in quantization: improving particular object retrieval in large scale image databases. In: IEEE conference on computer vision and pattern recognition, pp 1–8, June 2008
47. Marszałek M, Schmid C (2006) Spatial weighting for bag-of-features. In: IEEE conference on computer vision and pattern recognition, vol 2. pp 2118–2125
48. Cortes C, Vapnik V (1995) Support-vector networks. Mach Learn 20:273–297
49. Sreekanth V, Vedaldi A, Jawahar CV, Zisserman A (2010) Generalized RBF feature maps for efficient detection. In: Proceedings of the British machine vision conference (BMVC), 2010
50. Fathi A, Ren X, Rehg JM (2011) Learning to recognize objects in egocentric activities. In: The 24th IEEE conference on computer vision and pattern recognition, CVPR 2011, IEEE, Colorado Springs, 20–25 June 2011, pp 3281–3288
51. Over P, Awad G, Michel M, Fiscus J, Sanders G, Shaw B, Kraaij W, Smeaton AF, Quéenot G (2012) Trecvid 2012—an overview of the goals, tasks, data, evaluation mechanisms and metrics. In: Proceedings of TRECVID 2012, NIST, USA, 2012

Chapter 5
Evaluating Multimedia Features and Fusion for Example-Based Event Detection

Gregory K. Myers, Cees G. M. Snoek, Ramakant Nevatia,
Ramesh Nallapati, Julien van Hout, Stephanie Pancoast, Chen Sun,
Amirhossein Habibian, Dennis C. Koelma, Koen E. A. van de Sande
and Arnold W. M. Smeulders

G. K. Myers (✉) · J. van Hout · S. Pancoast
SRI International, 333 Ravenswood Avenue, Menlo Park, CA 94025, USA
e-mail: gregory.myers@sri.com

J. van Hout
e-mail: julien.vanhout@sri.com

S. Pancoast
e-mail: stephanie.pancoast@sri.com

C. G. M. Snoek · D. C. Koelma · K. E. A. van de Sande · A. W. M. Smeulders
University of Amsterdam (UvA), Science Park 904, P.O. Box 94323, 1098 GH Amsterdam,
The Netherlands
e-mail: cgmsnoek@uva.nl

D. C. Koelma
e-mail: koelma@uva.nl

K. E. A. van de Sande
e-mail: ksande@uva.nl

A. W. M. Smeulders
e-mail: ArnoldSmeulders@uva.nl

R. Nevatia · C. Sun
Institute for Robotics and Intelligent Systems, University of Southern California (USC),
Los Angeles, CA 90089-0273, USA
e-mail: nevatia@usc.edu

C. Sun
e-mail: chensun@usc.edu

R. Nallapati
Thomas J. Watson Research Center, 1101 Kitchawan Road, Yorktown Heights, NY 10598, USA
e-mail: nallapati@us.ibm.com

A. Habibian
University of Amsterdam (UvA)-Euvision, Matrix II, Science Park 400,
1098 XH Amsterdam, The Netherlands
e-mail: a.habibian@uva.nl

B. Ionescu et al. (eds.), *Fusion in Computer Vision*, Advances in Computer
Vision and Pattern Recognition, DOI: 10.1007/978-3-319-05696-8_5,
© Springer International Publishing Switzerland 2014

Abstract Multimedia event detection (MED) is a challenging problem because of the heterogeneous content and variable quality found in large collections of Internet videos. To study the value of multimedia features and fusion for representing and learning events from a set of example video clips, we created SESAME, a system for video SEarch with Speed and Accuracy for Multimedia Events. SESAME includes multiple bag-of-words event classifiers based on single data types: low-level visual, motion, and audio features; high-level semantic visual concepts; and automatic speech recognition (ASR). Event detection performance was evaluated for each event classifier. The performance of low-level visual and motion features was improved by the use of difference coding. The accuracy of the visual concepts was nearly as strong as that of the low-level visual features. Experiments with a number of fusion methods for combining the event detection scores from these classifiers revealed that simple fusion methods, such as arithmetic mean, perform as well as or better than other, more complex fusion methods.

5.1 Introduction

The goal of multimedia event detection (MED) is to detect user-defined events of interest in massive, continuously growing video collections, such as those found on the Internet. This is an extremely challenging problem because the contents of the videos in these collections are completely unconstrained, and the collections include user-generated videos. The quality of such videos varies widely, because they are often made with handheld cameras and may exhibit jerky motions, wildly varying fields of view, and poor lighting. The audio in these videos is recorded in a variety of acoustic environments, often with a single camera-mounted microphone, with no attempt to prevent background sounds from masking speech.

For purposes of this research, an event, as defined in the TREC Video Retrieval Evaluation (TRECVID) MED evaluation task sponsored by the National Institute of Standards and Technology (NIST) [1], has the following characteristics:

- It includes a complex activity occurring at a specific place and time.
- It involves people interacting with other people and/or objects.
- It consists of a number of human actions, processes, and activities that are loosely or tightly organized and have significant temporal and semantic relationships to the overarching activity.
- It is directly observable.

Figure 5.1 shows some sample video imagery from events in the TRECVID MED evaluation task. Events are more complex and may include actions (*hammering*, *pouring liquid*) and activities (*dancing*) occurring in different scenes (*street*, *kitchen*) in indoor and outdoor environments. Some events may be process-oriented, with an expected sequence of stages, actions, or activities (*making a sandwich* or *repairing an appliance*); other events may be a set of ongoing activities with no particular

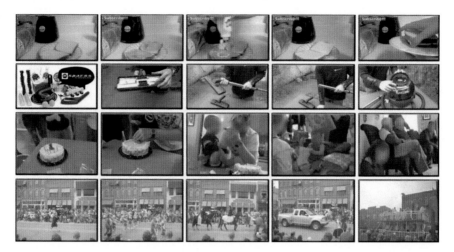

Fig. 5.1 Key frame series from example videos for the events *making a sandwich, repairing an appliance, birthday party,* and *parade* (The imagery was obtained from the Linguistic Data Consortium. Faces have been obscured for privacy)

beginning or end (*birthday party* or *parade*). An event may be observed in only a portion of the video clip, and relevant clips may contain extraneous content.

Multimedia event detection can be considered as a search problem with a query-retrieval paradigm. Currently, videos in online collections, such as YouTube, are retrieved based on text-based search. Text labels are either manually assigned when the video is added to the collection or derived from text already associated with the video, such as text content that occurs near the video in a multimedia blog or web page. Videos are searched and retrieved by matching a text-based user query to videos' text labels, but performance will depend on the quality and availability of such labels.

Highly accurate text-based video retrieval requires the text-based queries to be comprehensive and specific. In the TRECVID MED [1] evaluation, each event is defined by an "**event kit**," which includes a 150–400 word text description consisting of an event name, definition, explication (textual exposition of the terms and concepts), and lists of scenes, objects, people, activities, and sounds that would indicate the presence of the event. Figure 5.2 shows an example for the event *working on a woodworking project*. The user might also have to specify how similar events are distinguished from the event of interest (e.g., *not construction* in Fig. 5.2), and may have to estimate the frequency with which various entities occur in the event (e.g., *often indoors*). Subcategories and variations of the event may also have to be considered (e.g., operating a lathe in a factory).

The work described in this chapter focused on evaluating the various data types and fusion methods for MED. The remainder of the paper is organized as follows. A short state-of-the art is presented in Sect. 5.2. Our approach for example-based

> **Event name:** Working on a woodworking project
>
> **Definition:** One or more people fashion an object out of wood
>
> **Explication:** Woodworking is a popular hobby that involves crafting an object out of wood. Typical woodworking projects may range from creating large pieces of furniture to small decorative items or toys. The process for making objects out of wood can include cutting wood into smaller pieces with hand or machine tools, carving wood to shape it, sanding wood to smooth it, gluing wood pieces together, drilling holes into wood, and applying decorative finishes to the completed object. Woodworking is distinguished from construction, which typically involves creation of large, permanent structures such as houses, sheds, or buildings, which may or may not be made of wood.
>
> **Evidential description:**
>   **scene:** often indoors in a workshop, garage, artificial lighting, occasionally outdoors
>   **objects/people:** woodworking tools (automatic or non-automatic saws, sander, knife), paint, stains, sawhorses, toolbox, safety goggles
>   **activities:** cutting and shaping wood, attaching pieces of wood together, smoothing/sanding wood
>   **audio:** sounds from power tools, hand tools being used (hammer, saw, etc.); narration of the process

Fig. 5.2 Event kit for *working on a woodworking project*

MED, including methods for content extraction and fusion, is described in Sect. 5.3. Experimental results are described in Sect. 5.4, and Sect. 5.5 contains a summary and discussion.

5.2 Related Work and Motivation for Our Approach

Another approach to detect events is to define the event in terms of a set of example videos, which we call an example-based approach. Example videos are matched to videos in the collection using the same internal representation for each. In this approach, the system automatically learns a model of the event based on a set of positive and negative examples, taking advantage of well-established capabilities in machine learning and computer vision. This chapter considers an example-based approach with both nonsemantic and semantic representations.

Current approaches for MED [2–7] rely heavily on kernel-based classifier methods that use low-level features computed directly from the multimedia data. These classifiers learn a mapping between the computed features and the category of event that occurs in the video. Videos and events are typically represented as "bag-of-words" models composed of histograms of descriptors for each feature type, including visual, motion, and audio features. Although the performance of these models is quite effective, individual low-level features do not correspond directly to terms with semantic meaning, and therefore cannot provide human-understandable evidence of why a video was selected by the MED system as a positive instance of a specific event.

A second representation is in terms of higher-level semantic concepts, which are automatically detected in the video content [8–11]. The detectors are related to objects, like a *flag*; scenes, like a *beach*; people, like *female*; and actions, like *dancing*. The presence of concepts such as these creates an understanding of the content. However, except for a few entities such as faces, most individual concept detectors are not yet reliable [12]. Also, training detectors for each concept require annotated data, which usually involves significant manual effort to generate. In the future, it is expected that more annotated data sets will be available, and weakly supervised learning methods will help improve the efficiency of generating them. Event representations based on high-level concepts have started to appear in the literature [13–16].

For an example-based approach, the central research issue is to find an event representation in terms of the elements of the video that permits the accurate detection of the events. In our approach, an event is modeled as a set of multiple bags-of-words, each based on a single data type. Partitioning the representation by data type permits the descriptors for each data type to be optimized independently. Specific multimodal combinations of features, such as bimodal audiovisual features [3], can be considered a single data type within this architecture. To characterize the video content as comprehensively as possible, the data types we used included a set of heterogeneous low-level features (visual appearance, motion, and audio) and higher-level semantic concepts (visual concepts). We also used automatic speech recognition (ASR) to generate a bag-of-words model in which semantic concepts were expressed directly by words in the recognized speech. The resulting event model combined multiple sources of information from multiple data types and multiple levels of information.

As part of the optimization process for the low-level features, we investigated the use of difference coding techniques in addition to conventional coding methods. Because the information captured by difference coding is somewhat complementary to the information produced by the traditional bag-of-words, we anticipated an improvement in performance. We conducted experiments to compare the performance of difference coding techniques with conventional feature coding techniques.

The remaining challenge is finding the best method for combining the multiple bags-of-words in the event-detection decision process. In the computer vision and multimedia retrieval literature, several fusion methods have been explored [3, 5, 17–19]. For event detection, the most common approach is to apply late fusion methods [3, 5, 17] in which the results for each data type are combined by fusing the decision scores from multiple event classifiers. This is a straightforward way of using the information from all data types in proportion to their relative contribution to event detection on videos with widely diverse content. We evaluated the performance of several late fusion methods.

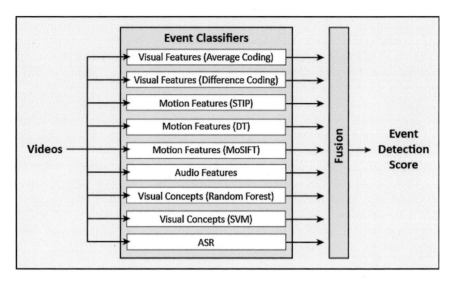

Fig. 5.3 Major components of the SESAME system

All of the experiments for evaluating the performance of the MED capability were performed using the data provided in the TRECVID MED [1] evaluation task. The MED evaluation uses the Heterogeneous Audio Visual Internet Collection (HAVIC) video data collection [20], which is a large corpus of Internet multimedia files collected by the Linguistic Data Consortium.

5.3 Approach for Example-Based MED

The work in this chapter focuses on SESAME, an MED system in which an event is specified as a set of video clip examples. A supervised learning process trains an event model from positive and negative examples, and an event classifier uses the event model to detect the targeted event. An event classifier was built for each data type. The results of all the event classifiers were then combined by fusing their decision scores. An overview of the SESAME system and methods for event classification and fusion are described in the following sections.

5.3.1 SESAME System Overview

The major components of the SESAME system are shown in Fig. 5.3. A total of nine event classifiers generate event detection decision scores: two based on low-level visual features, three based on low-level motion features, one based on low-level audio features, two based on visual concepts, and one based on ASR. (The particular

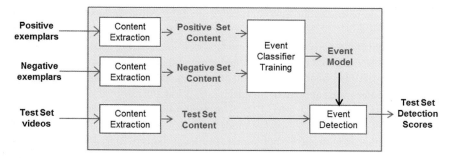

Fig. 5.4 Example-based event classifier for MED

set of classifiers used for each feature type was experimentally determined to be optimal with respect to performance on a reduced dataset.) The outputs of the event classifiers are combined by the fusion process.

Figure 5.4 shows the processing blocks within each event classifier. Each event classifier operates on a single type of data and includes both training and event classification. Content is extracted from positive and negative video examples, and the event classifier is trained, resulting in an event model. The event model produces event detection scores when it is applied to a test set of videos. Figure 5.4 does not show off-line training and testing to optimize the parameter settings for the content extraction processes.

5.3.2 Content Extraction Methods

This section describes the feature coding and aggregation methods that were common to the low-level features and the content extraction methods for the different data types: low-level visual features, low-level motion features, low-level audio features, high-level visual features, and ASR.

5.3.2.1 Feature Coding and Aggregation

The coding and aggregation of low-level features share common elements that we describe here. We extracted local features and aggregated them by using three approaches: conventional bag of words (BOW), vector of locally aggregated descriptors (VLAD), and Fisher vectors (FV).

The conventional BOW approach partitions low-level features into clusters to generate a codebook. Given a set of features from a video, a histogram is generated by assigning each feature from the set to one or several nearest code words. Several modifications to this approach are possible. One variation uses soft coding, where instead of assigning each feature to a single code word, distances from the code words are used to weigh the histogram terms for the code words. Another variation describes code words by a Gaussian mixture model (GMM), rather than just by the center of a cluster.

While conventional BOW aggregation has been successfully used for many applications, it does not maintain any information about the distribution of features in the feature space. FV has been introduced in previous work [21] to capture more detailed statistics, and has been applied to image classification and retrieval [22, 23] and capturing variation in time in video [24]. The basic idea is to represent a set of data by a gradient of its log-likelihood to model parameters and to measure the distance between instances with the Fisher kernel. For local features extracted from videos, it becomes natural to model their distribution as GMMs, forming a soft codebook. With GMM, the dimension of FV is linear in the number of mixtures and local feature dimensions.

Finally, VLAD [22] is proposed as a nonprobabilistic version of FV. It uses k-means instead of GMM, and accumulates the relative positions of feature points to their single nearest neighbors in the codebook.

Compared with conventional BOW, FV and VLAD have the following benefits:

- FV takes GMM as the underlying generative model.
- Both FV and VLAD are derivatives, so feature points with the same distribution as the general model has no overall impact on the video-level descriptors; as a result, FV and VLAD can suppress noisy and redundant signals.

None of the above aggregation methods consider feature localization in space or in time. We introduced a limited amount of this information by dividing the video into temporal segments (for time localization) and spatial pyramids (for spatial localization). We then compute the features in each segment or block separately and concatenate the resulting features. The spatial pooling and temporal segmentation parameters that yielded the best performance were determined through experimentation.

5.3.2.2 Visual Features

Two event classifiers were developed based on low-level visual features that have proven themselves for general video categorization in TRECVID [25]. They both follow a pipeline consisting of four stages: spatiotemporal sampling of points of interest, visual description of those points, encoding the descriptors into visual words, and supervised learning with kernel machines.

Spatiotemporal Sampling: The visual appearance of an event in video may have a dependency on the spatiotemporal viewpoint under which it is recorded. Salient point methods [26] introduce robustness against viewpoint changes by selecting points, which can be recovered under different perspectives. To determine salient points, Harris-Laplace relies on a Harris corner detector; applying it on multiple scales makes it possible to select the characteristic scale of a local corner using the Laplacian operator. For each corner, the Harris-Laplace detector selects a scale-invariant point if the local image structure under a Laplacian operator has a stable maximum.

Another solution is to use many points by dense sampling. For imagery with many homogenous areas, such as outdoor snow scenes, corners may be rare, so relying on a Harris-Laplace detector can be suboptimal. To counter the shortcomings of Harris-Laplace, we used dense sampling, which samples an image grid in a uniform fashion, using a fixed pixel interval between regions.

In our experiments, we used an interval distance of six pixels and sampled at multiple scales. Appearance variations caused by temporal effects were addressed by analyzing video beyond the key frame level [27]. Taking more frames into account during analysis allowed us to recognize events that were visible during the video, but not necessarily in a single key frame. We sampled one frame every two seconds. Both Harris-Laplace and dense sampling give an equal weight to all keypoints, regardless of their spatial location in the image frame. To overcome this limitation, Lazebnik et al. [28] suggest repeated sampling of fixed subregions of an image, e.g., 1×1, 2×2, 4×4, etc., and then aggregating the different resolutions into a spatial pyramid, which allows for region-specific weighting. Since every region is an image in itself, the spatial pyramid can be combined with both the Harris-Laplace point detector and dense point sampling. We used a spatial pyramid of 1×1 and 1×3 regions, because this was the set of regions that yielded the best performance in our experiments.

Visual Descriptors: In addition to the visual appearance of events in the spatiotemporal viewpoint under which they are recorded, the lighting conditions during recording also play an important role in MED. Properties of color features under classes of illumination and viewing features, such as viewpoint, light intensity, light direction, and light color, can change, specifically for real-world datasets as considered within TRECVID [29]. We followed [25] and used a mixture of SIFT, OpponentSIFT, and C-SIFT descriptors. The SIFT feature proposed by Lowe [30] describes the local contrast of a region using edge-orientation histograms. Because the SIFT feature is normalized, the gradient magnitude changes have no effect on the final feature. OpponentSIFT describes all the channels in the opponent color space using SIFT features. The information in the O3 channel is equal to the intensity information, while the other channels describe the color information in the image. The feature normalization, as effective in SIFT, cancels out any local changes in light intensity. In the opponent color space, the O1 and O2 channels still contain some intensity information. To add invariance to shadow and shading effects, the C-invariant [31] eliminates the remaining intensity information from these channels. The C-SIFT feature uses the C-invariant, which can be seen as the gradient (or derivative) for the normalized opponent color space O1/I and O2/I. The I intensity channel remains unchanged. C-SIFT is known to be scale-invariant with respect to light intensity. We computed the SIFT and C-SIFT descriptors around salient points obtained from the Harris-Laplace detector and dense sampling. We then reduced all descriptors to 80 dimensions with principal component analysis (PCA), a common procedure in the video categorization literature [25].

Word Encoding: To avoid using all low-level visual features from a video, we followed the well-known codebook approach. We first assigned the features to discrete codewords from a predefined codebook. Then, we used the frequency distribution of the codewords as a compact feature vector representing an image frame. Based on [25], we employed codebook construction using k-means clustering in combination with average codeword assignment and a maximum of 4,096 codewords. (The number of codewords and the values of other parameters selected for this approach were determined through experimentation.) The traditional hard assignment can be improved by using soft assignment through kernel codebooks [32]. A kernel codebook uses a kernel function to smooth the hard assignment of (image) features to codewords by assigning descriptors to multiple clusters weighted by their distance to the center. We also used difference coding, with VLAD performing k-means clustering of the PCA-reduced descriptor space with 1,024 components. The output of the word encoding is a BOW vector using either hard average coding or soft VLAD coding. The BOW vector forms the foundation for event detection.

Kernel Learning: Kernel-based learning methods are typically used to develop robust event detectors from audiovisual features. As described in [25], we relied predominantly on the support vector machine framework for supervised learning of events: specifically, the LIBSVM[1] implementation with probabilistic output. To handle imbalance in the number of positive versus negative training examples, we fixed the weights of the positive and negative classes by estimating the prior probabilities of the classes on training data. We used the histogram intersection kernel and its efficient approximation as suggested by Maji et al. [33]. For difference coded BOWs, we used a linear kernel [21].

Experiments: We evaluated the performance of these two event classifiers on a set of 12,862 drawn from the training and development data from the TRECVID MED [1] evaluation. This SESAME Evaluation dataset consisted of a training set of 8,428 videos and a test set of 4,434 videos sampled from 20 event classes and other classes that did not belong to any of the 20 events. To make good use of the limited number of available positive instances of events, the positives were distributed so that, for each event, there were approximately twice as many positives in the training set as there were in the test set. Separate classifiers were trained for each event based on a one-versus-all paradigm. Table 5.1 shows the performance of the two event classifiers measured by mean average precision (MAP). Color-average coding with a histogram intersection kernel (HIK) Support Vector Machine (SVM) slightly outperformed color-difference soft coding with a linear SVM. For events such as *changing a vehicle tire* and *town hall meeting*, the average HIK was the best event representation. However, for some events, such as *flash mob gathering* and *dog show*, the difference coding was more effective. To study whether the representations complement each other, we also performed a simple average fusion; the results indicate a further increase in event detection performance, improving MAP from 0.342 to 0.358 and giving the best overall performance for the majority of events.

---

[1] http://www.csie.ntu.edu.tw/~cjlin/libsvm/

Table 5.1 Mean average precision (MAP) **of event classifiers with** low-level visual features and their fusion for 20 TRECVID MED [1] evaluation event classes

Event[a]	Average coding with HIK SVM	Difference coding with linear SVM	Fusion
Birthday_party	**0.275**	0.229	0.261
Changing_a_vehicle_tire	**0.305**	0.269	0.302
Flash_mob_gathering	0.602	**0.644**	0.636
Getting_a_vehicle_unstuck	0.457	**0.496**	0.494
Grooming_an_animal	**0.280**	0.222	0.275
Making_a_sandwich	0.268	0.278	**0.314**
Parade	0.416	0.415	**0.427**
Parkour	**0.464**	0.413	0.450
Repairing_an_appliance	0.486	0.469	**0.498**
Working_on_a_sewing_project	0.378	0.388	**0.400**
Attempting_a_bike_trick	0.398	0.350	**0.408**
Cleaning_an_appliance	**0.138**	0.077	0.135
Dog_show	0.595	**0.651**	0.636
Giving_directions_to_a_location	0.123	0.130	0.134
Marriage_proposal	0.058	**0.093**	0.071
Renovating_a_home	0.229	0.273	**0.285**
Rock_climbing	0.488	0.466	**0.507**
Town_hall_meeting	**0.531**	0.463	0.502
Winning_a_race_without_a_vehicle	0.237	**0.284**	0.263
Working_on_a_metal_crafts_project	0.109	0.134	**0.153**
Mean for all events	**0.342**	0.337	0.358

[a] Best result per event is denoted in bold

5.3.2.3 Motion Features

Many motion features for activity recognition have been suggested in previous work; [4] provides a nice evaluation of motion features for classifying web videos on the NIST MED 2011 dataset. Based on our analysis of previous work and some small-scale experiments, we decided to use three features: spatio-temporal interest points (STIPs), dense trajectories (DTs) [34], and MoSIFT [35]. STIP features are computed at corner-like locations in the 3-D spatio-temporal volume. Descriptors consist of histograms of gradient and optical flow at these points. This is a very commonly used descriptor; more details may be found in [36]. Dense trajectory features are computed on a dense set of local trajectories (typically computed over 15 frames). Each trajectory is described by its shape and by histograms of intensity gradient, optical flow, and motion boundaries around it. Motion boundary features are somewhat invariant to camera motion. MoSIFT, as its name suggests, uses SIFT feature descriptors; its feature detector is built on motion saliency. STIP and DT were extracted using the default parameters as provided[2]; the MoSIFT features were obtained in the form of coded BOW features.[3]

After the extraction of low-level motion features, we generated a fixed-length video-level descriptor for each video. We experimented with the coding schemes described in Sect. 5.3.2.1 for the STIP and DT features; for MoSIFT, we were able to use BOW features only. We used the training and test sets described above.

We trained separate SVM classifiers for each event and each feature type. Training was based on a one-versus- all paradigm. For conventional BOW features, we used the χ^2 kernel. We used the Gaussian kernel for VLAD and FV. To select classifier-independent parameters (such as the codebook size), we conducted fivefold cross validation of 2,062 videos from 15 event classes. We conducted fivefold cross validation on the training set to select classifier-dependent parameters. For BOW features, we used 1,000 codewords; for FV and VLAD, we used 64 cluster centers. More details of the procedure are found in [37].

We compared the performance of conventional BOW, FV, and VLAD for STIP features; BOW and FV for DT features; and BOW for MoSIFT, using the SESAME Evaluation dataset. Table 5.2 shows the results.

We can see that FV gave the best MAP for both STIP and DT. VLAD also improved MAP for STIP, but was not as effective as the FV features. We were not able to perform VLAD and FV experiments for MoSIFT features, but would expect to have seen similar improvements there.

5.3.2.4 Audio Features

The audio is modeled as a *bag of audio words* (BOAW). The BOAW has recently been used for audio document retrieval [38] and copy detection [39], as well as MED tasks [40]. Our recent work [41] describes the basic BOAW approach. We extracted the audio data from the video files and converted them to a 16 kHz sampling rate. We extracted Mel frequency cepstral coefficients (MFCCs) for every 10 ms interval using a hamming window with 50 % overlap. The features consist of 13 values (12 coefficients and the log-energy), along with their delta and delta-delta values. We used a randomized sample of the videos from the TRECVID 2011 MED evaluation development set to generate the codebook. We performed k-means clustering on the MFCC features to generate 1,000 clusters. The centroid for each cluster is taken as a code word. The soft quantization process used the codebook to map the MFCCs to code words. We trained an SVM classifier with a histogram intersection kernel on the soft quantization histogram vectors of the video examples, and used the classifier to detect the events. Evaluation with the SESAME Evaluation dataset showed that the audio features achieved a MAP of 0.112.

---

[2] We obtained the STIP code from http://www.di.ens.fr/~laptev/download/stip-1.1-winlinux.zip, and DT code from http://lear.inrialpes.fr/people/wang/dense_trajectories.

[3] MoSIFT features were provided by Dr. Alex Hauptmann of Carnegie-Mellon University.

Table 5.2 Mean average precision of event classifiers with motion features for 20 TRECVID MED [1] evaluation event classes

Event[a]	BOW + MoSIFT	BOW + STIP	VLAD + STIP	FV + STIP	BOW + DT	FV + DT
Birthday_party	0.191	0.217	0.217	0.189	0.225	**0.293**
Changing_a_vehicle_tire	0.126	0.064	0.165	0.136	0.190	**0.217**
Flash_mob_gathering	0.463	0.535	**0.579**	0.569	0.564	0.567
Getting_a_vehicle_unstuck	0.337	0.284	0.316	0.365	0.403	**0.439**
Grooming_an_animal	**0.290**	0.093	0.116	0.147	0.216	0.247
Making_a_sandwich	0.164	0.154	0.193	0.225	0.198	**0.234**
Parade	0.326	0.260	0.364	**0.457**	0.446	0.419
Parkour	0.295	0.366	0.404	0.369	0.413	**0.459**
Repairing_an_appliance	0.368	0.357	0.370	0.385	0.417	**0.443**
Working_on_a_sewing_project	0.270	0.292	0.346	0.386	0.352	**0.433**
Attempting_a_bike_trick	**0.640**	0.104	0.234	0.235	0.245	0.438
Cleaning_an_appliance	**0.090**	0.058	0.088	0.074	0.066	0.089
Dog_show	0.488	0.361	0.489	0.557	0.600	**0.632**
Giving_directions_to_a_location	0.085	**0.194**	0.148	0.191	0.069	0.052
Marriage_proposal	0.027	0.040	0.107	**0.173**	0.059	0.118
Renovating_a_home	0.157	0.182	0.201	0.255	0.277	**0.361**
Rock_climbing	0.465	0.156	0.326	0.352	**0.470**	0.425
Town_hall_meeting	**0.519**	0.285	0.286	0.462	0.317	0.370
Winning_a_race_without_a_vehicle	**0.273**	0.187	0.174	0.260	0.179	0.216
Working_on_a_metal_crafts_project	0.116	**0.148**	0.064	0.032	0.072	0.128
MAP	**0.285**	0.217	0.259	0.291	0.289	0.329

[a] Best result per event is denoted in bold

5.3.2.5 Visual Concepts

Two event classifiers were based on concept detectors. We followed the pipeline proposed in [42]. We decoded the videos by uniformly extracting one frame every 2 sec. We then applied all available concept detectors to the extracted frames. After we concatenated the detector outputs, each frame was represented by a concept vector. Finally, we aggregated the frame representations into a video-level representation by averaging and normalization. On top of this concept representation per video, we used either a HIK SVM or a random forest as an event classifier.

To create the concept representation, we needed a comprehensive pool of concept detectors. We built this pool of detectors using the human-annotated training data from two publicly available resources: the TRECVID 2012 Semantic Indexing task [43] and the ImageNet Large-Scale Visual Recognition Challenge 2011 [44]. The former has annotations for 346 semantic concepts on 400,000 keyframes from web videos. The latter has annotations for 1,000 semantic concepts on 1300,000 photos. The categories are quite diverse and include concepts from various types; i.e., objects like *helicopter* and *harmonica*, scenes like *kitchen* and *hospital*, and actions like *greeting* and *swimming*. Leveraging the annotated data available in these datasets, we trained 1,346 concept detectors in total.

Table 5.3 Mean average precision of event classifiers with visual concept features for 20 TRECVID MED [1] evaluation event classes

Event[a]	RF	SVM
Birthday_party	**0.339**	0.324
Changing_a_vehicle_tire	**0.251**	0.241
Flash_mob_gathering	0.542	0.542
Getting_a_vehicle_unstuck	**0.454**	0.426
Grooming_an_animal	**0.254**	0.231
Making_a_sandwich	**0.283**	0.257
Parade	**0.373**	0.306
Parkour	**0.550**	0.479
Repairing_an_appliance	**0.422**	0.404
Working_on_a_sewing_project	0.390	**0.394**
Attempting_a_bike_trick	**0.475**	0.472
Cleaning_an_appliance	0.097	**0.149**
Dog_show	**0.595**	0.529
Giving_directions_to_a_location	0.058	**0.097**
Marriage_proposal	**0.077**	0.066
Renovating_a_home	0.295	**0.325**
Rock_climbing	**0.412**	0.401
Town_hall_meeting	0.411	**0.417**
Winning_a_race_without_a_vehicle	**0.198**	0.167
Working_on_a_metal_crafts_project	0.099	**0.162**
Mean for all events	**0.341**	0.330

[a] Best result per event is denoted in bold

We followed the state-of-the-art for our implementation of the concept detectors. We used densely sampled SIFT, OpponentSIFT, and C-SIFT descriptors, as we had for our event detector using visual features, but this time, we used difference coding with Fisher vectors [21]. We used a visual vocabulary of 256 words. We again used the full image and three horizontal bars as a spatial pyramid. The feature vectors representing the training images formed the input for a linear SVM.

Experiments with the SESAME Evaluation dataset, summarized in Table 5.3, show that the random forest classifier is more successful than the nonlinear HIK SVM for event detection using visual concepts, although the two approaches are quite close on average. Note that the event detection results using visual concepts are close to our low-level representation using visual or motion features.

5.3.2.6 Automatic Speech Recognition

Spoken language content is often present in user-generated videos and can potentially contribute useful information for detecting events. The recognized speech has direct semantic information that typically complements the information contributed by low-level visual features. We used DECIPHER, SRI's ASR software, to recognize spoken English. We used acoustic and language models obtained from an ASR system

[45] trained on speech data recorded in meetings with a far-field microphone. Initial tests on the audio in user-generated videos revealed that the segmentation process, which distinguishes speech from other audio, often misclassified music as speech. Therefore, before running the speech recognizer on these videos, we constructed a new segmenter, which is described below.

The existing segmenter was GMM-based and had two classes (speech and non-speech). For this effort, we leveraged the availability of annotated TRECVID video data and built a segmenter better tuned to audio conditions in user-generated videos. We built a segmenter with four classes: speech, music, noise, and pause. We measured the effectiveness of the new segmentation by the word-error rates (WERs) obtained by feeding the speech-segmented audio to our ASR system. We found that the new segmentation helped reduce the WER from 105 to 83 %. This confirmed that the new segmentation models were a better match to the TRECVID data than models trained on meeting data. For reference, when all the speech segments were processed by the ASR, the WER obtained by our system was 78 %. (This oracle segmentation provided the lowest WER that could be achieved by improving the segmentation.)

To create features for the event classifiers, we used ASR recognition lattices to compute the expected word counts for each word and each video. This approach provided significantly better results compared to using the 1-best ASR output, because it compensated for ASR errors by including words with lower posteriors that weren't necessarily present in the 1-best. We computed the logarithm of the counts for each word, appended them to form a feature vector of dimension 34,457, and used a linear SVM for the event classifiers. More details may be found in [46]. Evaluation with the SESAME Evaluation dataset showed that the ASR event classifiers achieved a MAP of 0.114.

5.3.3 Fusion

We implemented a number of late fusion methods, all of which involved a weighted average of the detection scores from the individual event classifiers. The methods for determining the weights considered several factors:

- **Event dependence and learned weights:** Because the set of most reliable data types for different events might vary, we considered the importance of learning the fusion weights for each event using a training set. However, when there is limited data available for training, aggregating the data for all events and computing a fixed set of weights for all events may yield more reliable results. Another strategy is to set the weights without training with any data at all. For example, in the method of fusing with the arithmetic mean of the scores, all of the weights are equal.
- **Score dependence:** For weights learned via cross-validation on a training set, a single set of fixed weights might be learned for the entire range of detection scores. Alternatively, the multidimensional space of detection scores might be partitioned into a set of regions, with a set of weights assigned to each region. In general, more data is needed for score-dependent weights to avoid overfitting.

- **Adjustment for missing scores:** When the scores for some types of data (particularly for ASR and MFCC) are missing, a default value, such as an average for the missing score, might be used, but this could provide a misleading indication of contribution and therefore degrade performance. Another way in which the fusion methods described below dealt with missing scores was renormalizing the weights of the nonmissing scores. Another alternative was learning multiple sets of weights, each set for a particular combination of nonmissing scores.

We evaluated the fusion models described below. All of the models operated on detection scores that were normalized using a Gaussian function (i.e., computing the z-score by removing the mean and scaling by the standard deviation).

Arithmetic mean (AM): In this method, we compute the AM of the scores of the observed data types for a given clip. Missing data types for a given clip are ignored, and the averaging is performed over the scores of observed data types.

Geometric mean (GM): In this method, we compute the uniform GM of the scores of the observed data types for a given clip. As we do for AM, we ignore missing data types and compute the geometric mean of the scores from observed data types.

Mean average precision-weighted fusion (MAP): This fusion method weighs scores from the observed data types for a clip by their normalized average precision scores, as computed on the training fold. Again, the normalization is performed only over the observed data types for a given clip.

Weighted mean root (WMR): This fusion method is a variant of the MAP-weighted method. In this method, we compute the fusion score as we do for MAP-weighted fusion, except the final fused score x' is determined by performing a power normalization of the MAP-based fused score x:

$$x' = x^{\frac{1}{\alpha}} \tag{5.1}$$

where α is the number of nonmissing data types for that video. In other words, the higher the number of data types from which the fusion score is computed, the more trustworthy the output.

Conditional mixture model: This model combines the detection scores from various data types using mixture weights that were trained by the expectation maximization (EM) algorithm on the labeled training folds. For clips that are missing scores from one or more data types, we provide the expected score for that data type based on the training data.

Sparse mixture model (SMM): This extension of the conditional mixture model addresses the problem of missing scores for a clip by computing a mixture for only the observed data types [47]. This is done by renormalizing the mixture weights over the observed data types for each clip. The training was done with the EM algorithm, but the maximization step no longer had a closed-form solution, so we used gradient-descent techniques to learn the optimal weights.

5 Evaluating Multimedia Features and Fusion for Example-Based Event Detection

Table 5.4 Fusion methods and their characteristics

Fusion method	Event-independent?	Learned on a training set?	Score-dependent?	Adjustment for missing scores?
Arithmetic mean	Yes	No	No	Yes
Geometric mean	Yes	No	No	Yes
MAP-weighted	No	Yes	No	Yes
Weighted mean root	No	Yes	No	Yes
Conditional mixture model	No	Yes	No	No
Sparse mixture model	No	Yes	No	Yes
SVMLight	No	Yes	Yes	No
Distance from threshold	No	Yes	Yes	No
Bin accuracy weighting	No	Yes	Yes	No

SVMLight: This fusion model consists of training an SVM using the scores from various data types as the features for each clip. Missing data types for a given clip are assigned zero scores. We used the SVMLight[4] implementation with linear kernels.

Distance from threshold: This is a weighted averaging method [3] that dynamically adjusts the weights of each data type for each video clip based on how far the score is from its decision threshold. If the detection score is near the threshold, the correct decision is presumed to be somewhat uncertain, and a lower weight is assigned. A detection score that is much greater or much lower than the threshold indicates that more confidence should be placed in the decision, and a higher weight is assigned.

Bin accuracy weighting: This method tries to address the problem of uneven distribution of detection scores in the training set. For each data type, the range of scores in the training fold is divided into bins with approximately equal counts per bin. During training, the accuracy of each bin is measured by computing the proportion of correctly classified videos whose scores fall within the bin. During testing, for each data type, the specific bin that the scores fall into is determined, and the corresponding bin accuracy scores for each data type are used as fusion weights.

Table 5.4 summarizes the fusion methods and their characteristics.

5.4 Experimental Results

We evaluated the performance of our SESAME system using the data provided in the TRECVID MED [1] evaluation task. Although the MED event kit contained both a text description and video examples for each event, the SESAME system implemented the example-based approach in which only the video examples were used for event detection training.

---

[4] http://svmlight.joachims.org/

5.4.1 Evaluation by Data Type

Table 5.5 lists results on the SESAME Evaluation dataset, which consisted of a training set of 8,428 videos and a test set of 4,434 videos, sampled from 20 event classes and other classes that did not belong to any of the 20 events. In terms of the performance of the various data types, the visual features were the strongest performers across all events. The accuracy of the visual concepts was nearly as strong as that of the low-level visual features. The motion features also showed strong performance. Although the performance of low-level audio features and ASR was significantly less, ASR had the highest performance for events containing a relatively large amount of speech content, including a number of instructional videos. The best scores for each event are distributed among all of the data types, indicating that fusion of these data should yield improved performance. Indeed, the AM fusion of the individual event classifiers, which is listed in the last column of Table 5.5, shows a significant boost in performance: a 33 % improvement over the best single data type.

5.4.2 Evaluation of Fusion Methods

We tested the late fusion methods described in Sect. 5.3.3 using the SESAME Evaluation dataset. For all our fusion experiments, we trained each event classifier on the training set, and executed the classifier on the test set to produce detection scores for each event. To produce legitimate fusion scores over the test set, we used tenfold cross validation, with random fold selection, to generate the detections, and then obtained a micro-averaged average precision over the resulting detections. The micro-averaged MAP was computed by averaging the average precision for each event. To gauge the stability of the fusion methods, we repeated this process 30 times and computed the macro average and standard deviation of the micro-averaged MAPs. Because the Arithmetic Mean and Geometric Mean methods are untrained, their micro-averaged MAPs will be the same regardless of fold selection; thus, the standard deviations for their micro-averaged MAPs are zero.

Table 5.6 shows the MED performance of various fusion methods. The comparison indicates that the simplest fusion methods, such as AM and GM, performed as well as or better than other, more complex fusion methods. Also note that most of the top-performing fusion methods (AM, GM, MAP, WMR, and SMM) adjusted their weights to accommodate missing scores.

5.4.3 Evaluation of MED Performance in TRECVID

As the SESAME team, we participated in the 2012 TRECVID MED [1] evaluation and submitted the detection results for a system configured nearly the same as that

Table 5.5 Experiment results in terms of mean average precision for individual event classifiers

Event[a]	Low-level visual features		Visual concept features		Motion features			Audio		Fusion
	SIFT-AVG	SIFT-DC	RF	SVM	STIP	DT	MOSIFT	MFCC	ASR	AM
Birthday_party	0.275	0.229	**0.339**	0.324	0.189	0.293	0.191	0.146	0.062	0.372
Changing_a_vehicle_tire	**0.305**	0.270	0.251	0.241	0.136	0.217	0.126	0.024	0.209	0.343
Flash_mob_gathering	0.603	**0.644**	0.542	0.542	0.569	0.567	0.463	0.139	0.017	0.644
Getting_a_vehicle_unstuck	0.457	**0.496**	0.454	0.426	0.365	0.439	0.337	0.040	0.011	0.586
Grooming_an_animal	0.280	0.222	0.254	0.231	0.147	0.247	**0.290**	0.038	0.024	0.352
Making_a_sandwich	0.267	0.278	0.283	0.257	0.225	0.234	0.164	0.038	**0.378**	0.392
Parade	0.416	0.414	0.373	0.306	**0.457**	0.419	0.326	0.119	0.013	0.578
Parkour	0.464	0.414	**0.550**	0.479	0.369	0.459	0.295	0.029	0.009	0.564
Repairing_an_appliance	0.486	0.469	0.422	0.404	0.385	0.443	0.368	0.449	**0.517**	0.591
Working_on_a_sewing_project	0.378	0.388	0.390	**0.394**	0.386	0.433	0.270	0.192	0.276	0.551
Attempting_a_bike_trick	0.398	0.350	0.475	0.472	0.235	0.438	**0.640**	0.019	0.003	0.703
Cleaning_an_appliance	0.138	0.077	0.097	**0.149**	0.074	0.089	0.090	0.050	0.144	0.174
Dog_show	0.591	**0.650**	0.595	0.529	0.557	0.632	0.488	0.183	0.002	0.672
Giving_directions_to_a_location	0.123	0.130	0.058	0.097	**0.191**	0.052	0.085	0.075	0.066	0.193
Marriage_proposal	0.057	0.093	0.077	0.066	**0.173**	0.118	0.027	0.044	0.010	0.179
Renovating_a_home	0.229	0.273	0.295	0.325	0.255	**0.361**	0.157	0.099	0.145	0.461
Rock_climbing	**0.488**	0.466	0.412	0.401	0.352	0.425	0.465	0.020	0.005	0.615
Town_hall_meeting	**0.531**	0.463	0.411	0.417	0.462	0.370	0.519	0.433	0.341	0.649
Winning_a_race_without_a_vehicle	0.237	**0.284**	0.198	0.167	0.260	0.216	0.273	0.074	0.005	0.295
Working_on_a_metal_crafts_project	0.109	0.133	0.099	**0.162**	0.032	0.128	0.116	0.024	0.044	0.209
Mean for all events	0.342	0.337	**0.341**	0.330	0.291	0.329	0.285	0.112	0.114	0.456

[a] The data type with the highest MAP score for each event is in bold. The AM fusion of the individual event classifiers is listed in the last column

Table 5.6 MED performance of fusion methods with all event classifiers

Fusion method	MacroMAP	Standard deviation
Arithmetic mean	0.456	0.0000
Geometric mean	0.456	0.0000
MAP-weighted	0.437	0.0006
Weighted Mean Root	0.451	0.0005
Conditional mixture model	0.403	0.0054
Sparse mixture model	0.443	0.0007
SVMLight	0.451	0.0036
Distance from threshold	0.407	0.0005
Bin accuracy weighting	0.401	0.0031

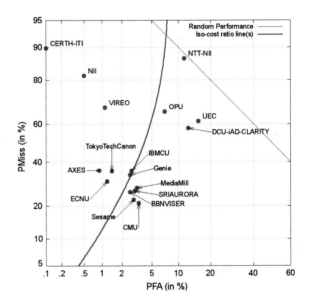

Fig. 5.5 Performance of the primary runs of 17 MED systems in the 2012 TRECVID MED [1] evaluation

described in this chapter.[5] The event classifiers were trained with all the positives from the event kit and negatives from the TRECVID MED training and development material. The test set consisted of the 99,000 videos used in the formal evaluation.

Figure 5.5 shows the performance of the primary runs of 17 MED systems in this evaluation in terms of miss and false alarm rates [48]. The performance of the SESAME run was one of the best among the evaluation participants.

---

[5] It included a poorer-performing ASR capability instead of the one described in Sect. 5.3.2.6, and a video OCR capability that contributed minimally to overall performance.

5.5 Summary and Discussion

SESAME, a MED capability that learns event models from a set of example video clips, includes a number of BOW event classifiers based on single data types: low-level visual, motion, and audio features; high-level semantic visual concepts; and ASR. Partitioning the representation by data type permits the descriptors for each data type to be optimized independently. We evaluated the detection performance for each event classifier and experimented with a number of fusion methods for combining the event detection scores from these classifiers. Our experiments using multiple data types and late fusion of their scores demonstrated strongly reliable MED performance.

Major conclusions from this effort include:

- The relative contribution of visual, motion, and audio features varies according to the specific event. This is due to differences in the relative distinctiveness and consistency of certain features for each event category. Across all events, score-level fusion resulted in a 33 % improvement over the best single data type, indicating that different types of features contribute to the representation of heterogeneous video data.
- The use of difference coding in low-level visual and motion features significantly improved performance. We surmise that difference coding works better than the traditional bag-of-words because it measures differences from the general model, which is likely to be dominated by the background features. We expect additional gains in performance if difference coding were applied to low-level audio features.
- The set of 1346 high-level visual features was nearly as effective as the set of low-level visual features. It appears that, in comparison to the 5000 or so concepts predicted to be needed for sufficient performance in event detection [49], this number of high-level features begins to span the space of concepts reasonably well. Therefore, analogous sets of motion and audio concepts should further improve overall performance.
- Although the performance of ASR was lower than that of the visual and motion features, its performance was highly event-dependent, and it performed reasonably well for events containing a relatively large amount of speech content, such as instructional videos.
- The simplest fusion methods for computing event detection scores were very effective compared to more complex fusion methods. One possible explanation for this is that the reliability of the scores is roughly equal across all data types. Another possible reason is that the limited number of positive training examples (an average of about 70 per event) is not enough to achieve the full benefit of the more complex fusion models.

While our relatively straightforward BOW approach was quite effective, we view it as a baseline capability that could be improved in several ways:

- Since the current approach aggregates low-level visual and motion features within fixed spatial partitions, the usage of local information is limited. Features of an object divided by our predefined partition, for example, will not be aggregated as a whole. We expect that the use of dynamic spatial pooling, which is better aligned to the structure and content of the video imagery, will improve performance. Segmenting the image into meaningful homogeneous regions would be even better, as it allows more salient characteristics to be extracted, and would eventually lead to better classification.
- The current approach ignores the temporal information within each video clip; all of the visual, motion, and audio features are aggregated. However, events consist of multiple components that appear at different times, so using time-based information for event modeling and detection should improve performance. Also, aggregating low-level features according to the temporal structure of the video may yield feature sets that better represent the video contents.
- All of the classifiers in our approach operate on a histogram of features and do not leverage any relationships between the features. Features occurring in video data are not generally independent. In particular, the combination of particular high-level semantic concepts could become strong discriminatory evidence, since their co-occurrence might be associated with a subset of relevant video content. For example, although the concepts *balloons* and *singing* occur in many contexts, the occurrence of both might be more common to *birthday party* than to other video content. Exploiting the spatiotemporal dependencies among the features would better characterize the video contents and offer a richer set of data with which to build event models.

Acknowledgments This work was supported by the Intelligence Advanced Research Projects Activity (IARPA) via the Department of Interior National Business Center, contract number D11PC0067. The U.S. Government is authorized to reproduce and distribute reprints for Governmental purposes notwithstanding any copyright annotation thereon. Disclaimer: The views and conclusions contained herein are those of the authors and should not be interpreted as necessarily representing the official policies or endorsements, either expressed or implied, of IARPA, DoI/NBC, or the U.S. Government.

References

1. Smeaton AF, Over P, Kraaij W (2006) Evaluation campaigns and TRECVID. In: Proceedings of the 8th ACM international workshop on multimedia information retrieval (MIR '06), Santa Barbara, 26–27 October 2006, ACM Press, New York, pp 321–330
2. Jiang Y-G, Bhattacharya S, Chang S-F, Shah M (2012) High-level event recognition in unconstrained videos. Int J Multimedia Inf Retrieval 2:1–29
3. Natarajan P, Wu S, Vitaladevuni S, Zhuang X, Tsakalidis S, Paurk U, Prasad R (2012) Multimodal feature fusion for robust event detection in web videos. In: Proceedings of the IEEE computer society conference on computer vision and pattern recognition (CVPR), pp 1298–1305

4. Sawhney H, Cheng H, Divakaran A, Javed O, Liu J, Yu Q, Ali S, Tamrakar A (2012) Evaluation of low-level features and their combinations for complex event detection in open source videos. In: CVPR, pp 2496–2499
5. Jiang Y (2013) Super: towards real-time event recognition in internet videos. In: ACM international conference on multimedia retrieval (ICMR), Article No. 33
6. Ballan L, Bertini M, Del Bimbo A, Seidenari L, Serra G (2011) Event detection and recognition for semantic annotation of video. Multimedia Tools Appl 51(1):279–302
7. Xu D, Chang S-F (2008) Video event recognition using kernel methods with multilevel temporal alignment. IEEE Trans Pattern Anal Mach Intell 30(11):1985–1997
8. Snoek CGM, Worring M (2009) Concept-based video retrieval. Found Trends Inf Retr 2(4):214–322
9. Felzenszwalb P, Girshick R, McAllester D, Ramanan D (2010) Object detection with discriminatively trained part-based models. IEEE TPAMI 32(9):1627–1645
10. Li L, Su H, Xing E, Fei-Fei L (2010) Object bank: a high-level image representation for scene classification and semantic feature sparsification. In: Advances in Neural Information Processing Systems, p. 24
11. Sadanand S, Corso JJ (2012) Action bank: a high-level representation of activity in video. In: CVPR
12. Snoek CGM, Smeulders AWM (2010) Visual-concept search solved? IEEE Comput 43(6):76–78
13. Merler M, Huang B, Xie L, Hua G, Natsev A (2012) Semantic model vectors for complex video event recognition. IEEE Trans Multimedia (TMM) 14(1):88–101
14. Althoff T, Song H, Darrell T (2012) Detection bank: an object detection based video representation for multimedia event recognition. In: ACM Multimedia (MM)
15. Tsampoulatidis I, Gkalelis N, Dimou A, Mezaris V, Kompatsiaris I (2011) High-level event detection in video exploiting discriminant concepts. In: Proceedings of the 1st ACM international conference on multimedia retrieval, p 8590
16. Habibian A, van de Sande KEA, Snoek CGM (2013) Recommendations for video event recognition using concept vocabularies. In: Proceedings of the ACM international conference on multimedia retrieval, Dallas, pp 89–96
17. Perera AGA, Oh S, Leotta M, Kim I, Byun B, Lee CH, McCloskey S, Liu J, Miller B, Huang ZF, Vahdat A, Yang W, Mori G, Tang K, Koller D, Fei-Fei L, Li K, Chen G, Corso J, Fu Y, Srihari R (2011) GENIE TRECVID 2011 multimedia event detection: late-fusion approaches to combine multiple audio-visual features. In: NIST TRECVID workshop
18. Pinquier J, Karaman S, Letoupin L, Guyot P, Mégret R, Benois-Pineau J, Gaëstel Y, Dartigues J-F (2012) Strategies for multiple feature fusion with Hierarchical HMM: application to activity recognition from wearable audiovisual sensors. In: International conference on pattern recognition, pp 3192–3195
19. Csurka G, Clinchant S (2012) An empirical study of fusion operators for multimodal image retrieval. In: Content-based multimedia indexing, pp 1–6
20. Strassel S, Morris A, Fiscus J, Caruso C, Lee H, Over P, Fiumara J, Shaw B, Antonishek B, Michel M (2012) Creating HAVIC: heterogeneous audio visual internet collection. In: Calzolari N, Choukri K, Declerck T, Uğur Doğan M, Maegaard B, Mariani J, Odijk J, Piperidis S (eds) Proceedings of the eighth international conference on language resources and evaluation, Istanbul
21. Jaakkola T, Haussler D (1999) Exploiting generative models in discriminative classifiers. In: Proceedings of the 1998 conference on advances in neural information processing systems II, pp 489–493
22. Jégou H, Perronnin F, Douze M, Sanchez J, Pérez P, Schmid C (2012) Aggregating local image descriptors into compact codes. IEEE TPAMI 34(9):1704–1716
23. Perronnin F, Dance C (2007) Fisher kernels on visual vocabularies for image categorization. In: CVPR
24. Mironica I, Uijlings J, Rostamzadeh N, Ionescu B, Sebe N (2013) Time matters!: capturing variation in time in video using fisher kernels. In: ACM multimedia, pp 701–704

25. Snoek CGM, van de Sande KEA, Habibian A, Kordumova S, Li Z, Mazloom M, Pintea SL, Tao R, Koelma DC, Smeulders AWM (2012) The MediaMill TRECVID 2012 semantic video search engine. In: Proceeding of TRECVID Workshop, Gaithersburg
26. Tuytelaars T, Mikolajczyk K (2008) Local invariant feature detectors: a survey. Found Trends Comp Graphics Vision 3(3):177–280
27. Snoek CGM, Worring M, Geusebroek J-M, Koelma DC, Seinstra FJ (2005) On the surplus value of semantic video analysis beyond the key frame. In: Proceedings of the IEEE international conference on multimedia and expo
28. Lazebnik S, Schmid C, Ponce J (2006) Beyond bags of features: spatial pyramid matching for recognizing natural scene categories. In: CVPR, New York, vol 2. pp 2169–2178
29. van de Sande KEA, Gevers T, Snoek CGM (2010) Evaluating color descriptors for object and scene recognition. IEEE TPAMI 32(9):1582–1596
30. Lowe DG (2004) Distinctive image features from scale-invariant keypoints. Int J Comput Vision 60:91–110
31. Geusebroek J-M, Boomgaard R, Smeulders AWM, Geerts H (2001) Color invariance. IEEE TPAMI 23(12):13381350
32. van Gemert JC, Snoek CGM, Veenman CJ, Smeulders AWM, Geusebroek J-M (2010) Comparing compact codebooks for visual categorization. Comput Vis Image Und 114(4):450–462
33. Maji S, Berg AC, Malik J (2008) Classification using intersection kernel support vector machines is efficient. In: Proceedings of the IEEE computer society conference on CVPR, Anchorage, pp 619–626
34. Wang H, Kläser A, Schmid C, Cheng-Lin L (2011) Action recognition by dense trajectories. In: CVPR, pp 3169–3176
35. Chen M-Y, Hauptmann A (2009) MoSIFT: recognizing human actions in surveillance videos. In: CMU-CS-09-161, Carnegie Mellon University
36. Laptev I (2005) On space-time interest points. Int J of Comput Vision 64(2/3):107–123
37. Sun C, Nevatia R (2013) Large scale web video classification by use of Fisher vectors. In: Workshop on applications of computer vision, Clearwater
38. Chechik G, Ie E, Rehn M, Bengio S, Lyon D (2008) Large-scale content-based audio retrieval from text queries. In: Proceedings of 1st ACM international conference on multimedia information retrieval (MIR '08), New York, pp 105–112
39. Uchida Y, Sakazawa S, Argawal M, Akbacak M (2010) KDDI labs and SRI international at TRECVID 2010: content-based copy detection. In: NIST TRECVID 2010 evaluation, workshop
40. Jiang Y, Zeng X, Ye G, Ellis D, Shah M, Chang S (2010) Columbia-UCF TRECVID 2010 multimedia event detection: combining multiple modalities, contextual concepts, and temporal matching. In: NIST TRECVID Workshop
41. Pancoast S, Akbacak M (2012) Bag-of-audio-words approach for multimedia event detection. In: Proceedings of interspeech
42. Merler M, Huang B, Xie L, Hua G, Natsev A (2012) Semantic model vectors for complex video event recognition. IEEE Trans Multimedia 14(1):88101
43. Over P, Awad G, Michel M, Fiscus J, Sanders G, Shaw B, Kraaij W, Smeaton AF, Quéenot G (2012) TRECVID 2012: an overview of the goals, tasks, data, evaluation mechanisms, and metrics. In: Proceedings of TRECVID. http://www-nlpir.nist.gov/projects/tvpubs/tv12.papers/tv12overview.pdf
44. Berg A, Deng J, Satheesh S, Su H, Li F-F (2011) Imagenet large scale visual recognition challenge. http://www.image-net.org/challenges/LSVRC/2011/
45. Janin A, Stolcke A, Anguera X, Boakye K, Çetin Ö, Frankel J, Zheng J (2006) The ICSI-SRI spring 2006 meeting recognition system. In: MLMI'06 proceedings of the third international conference on machine learning for multimodal, interaction, pp 444–456
46. van Hout J, Akbacak M, Castaneda D, Yeh E, Sanchez M (2013) Extracting audio and spoken concepts for multimedia event detection. In: International conference on acoustics, speech, and signal processing (ICASSP)

47. Nallapati R, Yeh E, Myers G (2012) Sparse mixture model: late fusion with missing scores for multimedia event detection. Algorithms and systems VII, SPIE multimedia content access
48. Fiscus J, Michel M (2012) TRECVID 2012 multimedia event detection task. In: NIST TRECVID 2012 evaluation, workshop
49. Hauptmann A, Yan R, Lin W-H, Christel M, Wactlar H (2007) Can high-level concepts fill the semantic gap in video retrieval? A case study with broadcast retrieval. IEEE Trans Multimedia 9(5):958–966

Chapter 6
Rotation-Based Ensemble Classifiers for High-Dimensional Data

Junshi Xia, Jocelyn Chanussot, Peijun Du and Xiyan He

Abstract In past 20 years, Multiple Classifier System (MCS) has shown great potential to improve the accuracy and reliability of pattern classification. In this chapter, we discuss the major issues of MCS, including MCS topology, classifier generation, and classifier combination, providing a summary of MCS applied to remote sensing image classification, especially in high-dimensional data. Furthermore, the recently rotation-based ensemble classifiers, which encourage both individual accuracy and diversity within the ensemble simultaneously, are presented to classify high-dimensional data, taking hyperspectral and multidate remote sensing images as examples. Rotation-based ensemble classifiers project the original data into a new feature space using feature extraction and subset selection methods to generate the diverse individual classifiers. Two classifiers: Decision Tree (DT) and Support Vector Machine (SVM), are selected as the base classifier. Unsupervised and supervised feature extraction methods are employed in the rotation-based ensemble classifiers. Experimental results demonstrated that rotation-based ensemble classifiers are superior to Bagging, AdaBoost and random-based ensemble classifiers.

J. Xia · J. Chanussot · X. He
GIPSA-lab, Grenoble Institute of Technology, 38400 Grenoble, France
e-mail: junshi.xia@gipsa-lab.grenoble-inp.fr

J. Chanussot
Faculty of Electrical and Computer Engineering, University of Iceland, Reykjavik, Iceland
e-mail: jocelyn.chanussot@gipsa-lab.grenoble-inp.fr

P. Du (✉)
Department of Geographical Information Science, Nanjing University, Nanjing 210093, China
e-mail: dupjrs@gmail.com

X. He
e-mail: xiyan.he@gipsa-lab.grenoble-inp.fr

6.1 Introduction

Learning from high-dimensional data has important applications in areas such as speech processing, medicine, monitoring urbanization using multisource images, mineralogy using hyperspectral images [34, 56, 62, 64, 75, 77, 93]. Despite constant improvements in computational learning algorithms, supervised classification of high-dimensional data is still a challenge largely due to the curse of dimensionality (Hughes phenomenon) [40]. This is because the training set is very limited when compared to the hundreds or thousands of dimensions in high-dimensional data [8, 54]. Big efforts on feature extraction and feature selection have been applied to the supervised classifiers [35, 59, 61, 67, 69, 83, 92]. Since each learning algorithm (feature selection/extraction, classifiers) has its own advantages and disadvantages, efficient methodologies have yet to be developed. One of the most usual ways to achieve that is Multiple Classifier System (MCS) [4, 7, 11, 20, 27, 48, 50, 74, 78, 79, 85].

MCS comes from the idea that seek advices from several persons to make the final decision, where the basic assumption is that combining the opinions will produce a decision that is better than the single opinion [48, 50, 74]. The individual classifiers (member classifiers) are constructed and their outputs are integrated according to a certain combination approach, to gain the final classification result. The outputs can be generated by the same classifier with different training sets, or by the different classifiers with same or different training set. The success of MCS not only depends on a set of appropriate classifiers, but also on the diversity within the ensemble, which referred to two conditions: accuracy and diversity [15, 47]. Accuracy requires a set of appropriate classifiers to be as accurate as possible. Diversity means the difference among the classification results. Combining similar classification results would not further improve the accuracy. Both theoretical and empirical studies demonstrated that using a good diversity measure is able to find the extent of diversity among classifier and estimate the improvement in accuracy of combining individual classifiers [50, 74]. However, Brown et al. pointed out that the diversity for classification tasks is still an ill-defined concept, and defining an appropriate diversity measure for MCS is still an open question [12].

Generally speaking, we often adapt three independent steps: topology selection, classifier generation, and classifier combination, to construct the MCS. In Sect. 6.2, we will give a review on the uses of MCS, including these steps and along with the application of remote sensing.

Rotation-based ensemble classifier is one of the current state-of-the-art ensemble classifier methods [72]. This algorithm constructs different training sets as follows: first, the feature set is divided into several disjoint sets on which the training set is projected. Second, the subtraining set is obtained from the projection results using bootstrapping technique. Third, feature extraction is used to rotate each obtained subtraining set. The components obtained from feature extraction are rearranged to form the dataset that is treated as the input of a single individual classifier. The final result is produced by combining the output of individual classifiers generated by repeating the above steps in multiple times.

In this chapter, we will apply rotation-based ensemble classifier to classify high-dimensional data. In particular, two classifiers: Decision Tree (DT) and Support Vector Machine (SVM), are selected as the base classifiers. Unsupervised and supervised feature extraction methods are employed to rotate the training set. The performances of rotation-based ensemble classifiers are evaluated by the high-dimensional remote sensing images.

The remainder of this article is organized as follows. In Sect. 6.2, we introduce the topology, classifier generation, and classifier combination approaches of MCS, summarize the advances of MCS to high-dimensional remote sensing data classification. The main idea and two implementations of rotation-based ensemble are shown in Sects. 6.3 and 6.4, respectively. Experimental results are presented in Sect. 6.5. The conclusion and perspective of this chapter are drawn in Sect. 6.6.

6.2 Multiple Classifier System

Different classifiers, such as parametric classifiers and non-parametric classifiers, have their own strengths and limitations. The famous 'no free lunch' theorem stated by Wolpert may be extrapolated to the point of saying that there is no single computational view that solves all problems [86]. In the remote sensing community, Giacinto et al. compared the performances of different classification approaches in various applications and found that no one could always gain the best result [32]. In order to alleviate this problem, MCS can provide the complementary information of the pattern classifiers and integrate the outputs of these pattern classifiers so as to make the best use of the advantages and bypass the disadvantages. Nowadays MCS are highlighted by review articles as a hot topic and promising trend in remote sensing image classification and change detection [4, 21].

Most of MCS approaches focus on integrating the supervised classifiers. Few works devote to combine unsupervised classification results, often called *cluster ensemble* [38, 41]. Gao et al. proposed an interesting work to combine multiple supervised and unsupervised models using graph-based consensus maximization [29]. Unsupervised models (clustering), which do not directly generate label prediction for each individual classifier, can provide useful constraints for the joint prediction of a set of related object. Thus, Gao et al. proposed to consolidate a classification solution by maximizing the consensus among both supervised predictions and unsupervised constraints based on the optimization problem on a bipartite graph [29]. Experimental results on three real applications demonstrate the benefits of the proposed method over existing alternatives. In this chapter, we focus on the combination of supervised classifiers.

The main issues of MCS design are [50, 74, 88]:

- MCS topology: How to interconnect individual classifiers.
- Classifier generation: How to generate and select valuable classifiers.
- Classifier combination: How to build a combination function which can exploit the strengths of the selected classifiers and combine them optimally.

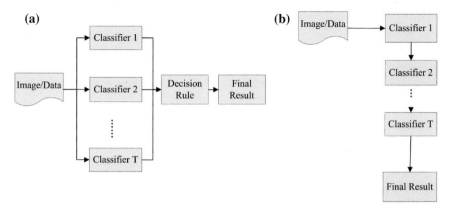

Fig. 6.1 The topologies of MCS. **a** Parallel style. **b** Concatenation style

6.2.1 MCS Topology

Figure 6.1 illustrates the two topologies employed in MCS design. The overwhelming majority of MCS reported in the literature is structured in a parallel style. In this architecture, multiple classifiers are designed independently without any mutual interaction and their outputs are combined according to certain strategies [70, 71, 90]. Alternatively, in the concatenation topology, the classification result generated by a classifier is used as the input into the next classifier [70, 71, 90]. When the primary classifier cannot obtain the satisfactory classification result, then the output of the primary classifier is feed to a secondary classifier, and so on. The main drawback of this topology is that the mistakes produced by the earlier classifier cannot be corrected by the later classifiers.

A very special case of concatenation topology is the AdaBoost [28]. The goal of AdaBoost is to enhance the accuracy of any given learning algorithm, even weak learning algorithms with an accuracy slightly better than chance. The algorithm processes training of the weak learner multiple times, each time presenting it with an updated weight over the training samples. Then, the weights of misclassified samples are increased to concentrate the learning algorithm on specific samples. Finally, the decisions generated by the weak learners are combined into a single decision.

6.2.2 Classifier Generation

Classifier generation aims to build mutually complementary individual classifiers that are accurate and at the same time disagree on some different parts of the input space. Diversity of individual classifiers is a vital requirement for the success of the MCS.

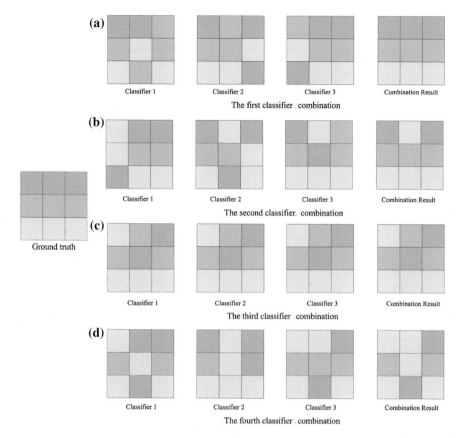

Fig. 6.2 Different classifier combinations using three single classifiers. The three colors represent the different classes. The overall accuracy of all individual classifier is 6/9. The overall accuracies of the four combinations are 1, 8/9, 6/9, and 5/9, respectively

Both theoretical and empirical studies indicate that we can ensure diversity using *Homogeneous* and *Heterogeneous* approaches [50, 74]. In *Homogeneous* approaches, we can obtain a set of classification results obtained by the same classifier by injecting randomness into the classifier, manipulating the training sample and the input features. The *Heterogeneous* approaches are to apply different learning algorithms to the same training set. First of all, we will start to review some diversity measures, and the generated classifiers followed to ensure the diversity in the ensemble.

6.2.2.1 Diversity Measures

Diversity represents the difference among the individual classifiers [15, 47]. Figure 6.2 presents four different classifier combinations within three classes (9 samples) using majority vote approach. Overall accuracy of each individual classifier is

Table 6.1 Summary of the 15 diversity measures

Name	p/n	s/dis/c	↑/↓	Range	Reference
Kappa statistic (κ_1, κ_2)	p	s	↓	[−1, 1]	[57, 60]
Mutual Information (MI)	p	s/c	↓	/	[43]
Q-statistic (Q)	p	s/c	↓	[−1, 1]	[49, 53]
Correlation coefficient (ρ)	p	s/c	↓	[−1, 1]	[53]
Double fault (DF)	p	s	↓	[0, 1]	[30]
Disagreement (Dis)	p	dis	↑	[0, 1]	[39]
Same fault (SF)	p	s	↓	[0, 1]	[2]
Weighted count of errors and correct (WCEC)	p	s	↑	/	[2]
Entropy (E)	n	s	↓	[0, 1]	[17, 53]
Kohavi-Wolpert variance (KW)	n	dis	↑	[0, 0.5]	[45]
Interrater agreement (IA)	n	s	↓	[0, 1]	[25, 53]
Generalized diversity (GD)	n	dis	↑	[0, 1]	[53, 68]
Conincident faiure diversity (CFD)	n	dis	↑	[0, 1]	[68]
Difficulty (θ)	n	dis	↓	[0, 0.25]	[37]

Note 'p' stands for 'pairwise' and 'n' stands for 'non pairwise', 's' means 'similarity,' 'c' means 'correlation' and 'dis' means 'dissimilarity.' The arrow specifies the greater diversity if the measure is lower (↓) or higher (↑)

6/9. The accuracies of the four combinations are 1, 8/9, 6/9, and 5/9, respectively. Our goal is to use diversity measures to find the classifier combination like in Fig. 6.2a or b and avoid to select the third or especially the fourth classifier combination.

Kuncheva and Whitaker summarized the diversity measures in classifier ensembles [53]. A special issue called "Diversity Measure in Multiple Classifier System" published in Information Fusion journal indicates that diversity measure is an important research direction in MCS [51]. Petrakos et al. applied agreement measure in decision fusion level combination [60]. Foody compared the different classification results from three aspects: similarity, non-inferiority and difference using hypothesis tests and confidence interval algorithms [26]. It is proved that increasing diversity should lead to better accuracy, but there is no formal proof of this dependency [12]. Table 6.1 summarizes the 15 diversity measures with their types, data range and literature sources.

Diversity measures also play an important role in ensemble pruning. Ensemble pruning aims at reducing the ensemble size prior to combination while maintaining a high diversity among the remaining members in order to reduce the computational cost and memory storage. To deal with the ensemble pruning process, several approaches have been proposed such as clustering-based, ranking-based, and optimization-based approaches [82].

6.2.2.2 Ensuring Diversity

Following the steps of pattern classification, we can enforce the diversity by the manipulation of training samples, features, outputs and classifiers.

Manipulating the training samples: In this method, each classifier is trained on different versions of training samples by exchanging the distribution of original training samples. This method is very useful for the unstable learner (decision tree and neural network), for which small changes in the training set will lead to a major change in the obtained classifier. Bagging and Boosting belong to this category [9, 28]. Bagging applies sampling with replacement to obtain the independent training samples for individual classifiers. Boosting changed the weights of training samples according to the results of the previous trained classifiers, focusing on the wrong classified samples, making the final result using a weight vote rule.

Manipulating the training features: The most well-known algorithm of this type is Random subspace [39]. Random subspace can be employed for several types of base learners, such as DT (Random Forest) [10], SVM [85]. Another development is Attribute Bagging, which establishes the appropriate size of a feature subsets, and then creates random projections of a given training set by random selection of feature subsets [13].

Manipulating the outputs: Multiclassification problem can be converted into several two-class classification problems. Each problem discover the discrimination between one class and the other classes. Error Correcting Output Coding (ECOC) adapts a code matrix to convert a multiclass problem into binary ones. Ensemble of multiclassifier classification problem can be treated as ensembles of multiple two-classifier classification problem, and then combined together [19]. The other method to deal with the outputs is label switching [58]. This method generates an ensemble by using perturbed version of the training set where the classes of the training samples are randomly switched. High accuracy can be achieved with fairly large ensembles generated by class switching.

Manipulating the individual classifiers: We can use different classifiers or the same classifier with different parameters to ensure the diversity. For instance, when the SVM is selected as the base learner, we can gain diversity by using different kernel functions or parameters.

6.2.3 Classifier Combination

Majority vote is a simple and an effective strategy for classifier combination. Within this scheme, a pixel is assigned as the class which gets the highest vote from the individual classifiers. Foody et al. used majority vote rule to integrate multiple binary classifiers for the mapping of a specific class [27]. According to the output of individual classifier, classifier combination approaches can be divided into three levels: abstract level, rank level, and measurement level [76]. The abstract level combination methods are applied when each classifiers outputs a unique label [76]. Rank level makes use of a ranked list of classes where ranking is based on decreasing likelihood. In the measurement level, probability values of the classes provided by each classifier are used in the combination. Majority/weighted vote, fuzzy integral, evidence theory, and dynamic classifier selection belong to the abstract level com-

Table 6.2 Summary of classifier combination approaches

Name	Hard labels	Soft labels	Validation set	Reference
Majority vote	Y	N	N	[50]
Weighted vote	Y	N	Y	[63, 90]
Bayesian average	N	Y	N	[30]
Dempster-shafer evidence theory	Y	N	Y	[50, 81]
Fuzzy integral	Y	N	Y	[46, 65]
Consensus theory	Y	Y	Y	[5, 6]
Dynamic classifier selection	Y	N	Y	[31, 78, 87]

Note "Y" and "N" mean whether or not the hard labels, soft labels or validation set are needed. Dynamic classifier selection method needs the original image to calculate the distance

bination methods. Bayesian average and Consensus theory belong to measurement level methods. Table 6.2 summarizes classifier combinational approaches. Weighted vote, fuzzy integral, Dempster-Shafer evidence theory and consensus theory require anther training set to calculate the weights. Dynamic classifier selection calculates the distance between the samples so it requires the original image. The computation time of dynamic classifier selection is more expensive than other approaches.

6.2.4 Applications to High-Dimensional Remote Sensing Data

Table 6.3 lists the studies of MCS applied to high-dimensional remote sensing images in recent years. These studies applied different effective MCS schemes to classify high-dimensional data, including multisource, multidate, and hyperspectral remote sensing data. In the works of Smits [78], Briem et al. [11], Gislason et al. [33], dynamic classifier selection, Bagging, Boosting and Random Forest are applied to classify multisource remote sensing data, respectively. Lawrence et al. [55], Kawaguchi and Nishii [44], Chan and Paelinckx [14], Rodriguez-Galiano et al. [73] used Boosting and Random Forest for the classification of multi-date remote sensing images. Doan and Foody [20] combining the soft classification results derived from NOAA AVHRR images using average operator and Evidence theory. From Table 6.2, the most well-known MCS approaches for hypespectral image classification is Random Forest. In Random Forest, each tree is trained on a bootstrapped sample of the original dataset and only a randomly chosen subset of the dimensions is considered for splitting a leaf. Thus, the computational complexity can be reduced and the correction between the trees are decreased. Apart from this, Waske et al. [85] developed random selection-based SVM for the classification of hyperspectral images. Yang et al. [91] proposed a novel subspace selection mechanism, dynamic subspace method, to improve random subspace method on automatically determining dimensionality and selecting component dimensions for diverse subspace. Du et al. [22] constructed diverse classifiers using different feature extraction methods and then combined the results using evidence theory, linear consensus algorithms. Recently,

6 Rotation-Based Ensemble Classifiers

Table 6.3 Studies on high-dimensional remote sensing image classification using MCS published in journals in recent years

Study	Methods	Datasets
Smits [78]	Dynamic classifier selection	Multispectral and SAR images
Briem et al. [11]	Bagging, Boosting and Consensus theory	Landsat MSS/AMSS+SAR and elevation, slope, aspect data
Lawrence et al. [55]	Stochastic gradient boosting	Multi-temporal Landsat TM images
Ham et al. [36]	Random Forest	Hyperspectral images
Gislason et al. [33]	Random Forest	Landsat MSS and elevation, slope, aspect data
Doan and Foody [20]	Average operator and Evidence theory	NOAA AVHRR images
Kawaguchi and Nishii [44]	AdaBoost with stump functions	Hyperspectral images
Chan and Paelinckx [14]	Random Forest and AdaBoost tree-based ensemble	Hyperspectral images
Waske et al. [84]	Random Forest	Hyperspectral images
Yang et al. [91]	Dynamic random subspace	Hyperspectral images
Waske et al. [85]	Random subspace	Hyperspectral images
Bakos and Gamba [3]	Hierarchical hybrid decision tree	Hyperspectral images
Du et al. [22]	Evidence theory, linear consensus	Hyperspectral images
Rodriguez-Galiano et al. [73]	Random Forest	Multi-temporal Landsat TM images
Xia et al. [89]	Rotation Forest	Hyperspectral images

Xia et al. [89] used Rotation Forest to classify hyperspectral remote sensing images. Compared to Random Forest, Rotation Forest [89] uses feature extraction to promote both the diversity and the accuracy of individual classifiers. Therefore, Rotation Forest can generate more accurate result than Random Forest.

6.3 Rotation-Based Ensemble Classifiers

In this study, rotation-based ensemble classifiers are used for high dimensional data. Let $\{\mathbf{X}, Y\} = \{\mathbf{x}_i, y_i\}_{i=1}^n$ be training samples. T is number of classifier. K is number of subsets (M: number of features in each subset). Γ is the base classifier. The details of rotation-based ensemble are presented in Algorithm 1 and Fig. 6.3 [66, 72]. According to Algorithm 1 and Fig. 6.3, the main steps of rotation-based ensemble classifier can be concluded as follows:

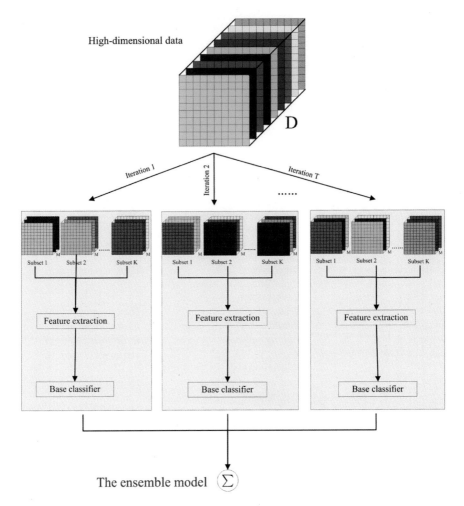

Fig. 6.3 Illustration of the rotation-based ensemble

- the input feature space is divided into K disjoint subspaces.
- feature extraction is performed on each subsets with the bootstrapped samples of 75 % size of $\{\mathbf{X}, Y\}$.
- the new training data, which is obtained by rotating the original training samples, is applied to the individual classifier.
- the individual classification results are combined using majority voting rule.

The strong performance is attributed to a simultaneous improvement of (1) diversity within the ensemble, obtained by the use of feature extraction on training data and (2) accuracy of the base classifiers, by keeping all extracted features in the training data [66, 72].

It is essential to notice step 5 in rotation-based ensemble presented in Algorithm 1, the sample size $\mathbf{X}'_{t,j}$ is selected smaller than $\mathbf{X}_{t,j}$ due to two reasons: one is to avoid obtaining the same coefficients when the same features are chosen and the other is to enhance the diversity within the ensemble [72].

Given the importance of the choice regarding the algorithm for feature extraction and the base classifier in rotation-based ensemble, several alternatives are considered in this study. The detailed feature extraction methods and base classifier can be found in the following section.

Algorithm 1 Rotation-based ensemble

Require: $\{\mathbf{X}, Y\} = \{\mathbf{x}_i, y_i\}_{i=1}^n$: training samples, T: number of classifier, K: number of subsets (M: number of features in each subset), Γ: base classifier
1. For $t = 1 : T$
2.   randomly split the features F into K subsets $F_{t,j}$
3.   For $j = 1 : K$
4.   select the corresponding features of $F_{t,j}$ to compose a new training features $\mathbf{X}_{t,j}$
5.   select a new training samples $\mathbf{X}'_{t,j}$ using bootstrap algorithm, whose size is 75 % of the original size
6.   transform $\mathbf{X}'_{t,j}$ to get the coefficients $v_{t,j}^{(1)}, \ldots, v_{t,j}^{(M_k)}$
7.   Endfor
8.   sparse matrix R_t is composed of the above coefficients

$$R_t = \begin{bmatrix} v_{t,1}^{(1)}, \ldots, v_{t,1}^{(M_1)} & 0 & \cdots & 0 \\ 0 & v_{t,2}^{(1)}, \ldots, v_{t,2}^{(M_2)} & \cdots & 0 \\ \vdots & \vdots & \ddots & \vdots \\ 0 & 0 & \cdots & v_{t,j}^{(1)}, \ldots, v_{t,j}^{(M_K)} \end{bmatrix}$$

9.   rearrange R_t to R_t^a with respect to the original feature set
10.   obtain the new training samples $\{\mathbf{X} R_t^a, Y\}$
11.   build classifier Γ_t using $\{\mathbf{X} R_t^a, Y\}$
12. Endfor
Ensure: the class label of given sample $\mathbf{x}$ predicted by multiple classifier
$\Gamma^*(\mathbf{x}) = \arg\max_{y \in \phi} \sum_{t=1}^T I\left(\Gamma_t\left(\mathbf{x} R_t^a\right) = y\right)$
$I(a = b)$ equals to 1 when a equals to b, otherwise equals to 0

6.4 Two Implementations of Rotation-Based Ensemble

6.4.1 Rotation Forest

Decision trees are often used for the multiple classifier system, especially for the rotation-based ensembles, because it is sensitive and fast. In this chapter, we adapt Classification and Regression Tree (CART) to construct Rotation Forest (RoF).

CART is a nonparametric decision tree learning technique, which can be both used for classification and regression. Decision trees are formed by a collection of rules based on variables in the modeling dataset: (1) rules based on variables's values are selected to get the best split to differentiate observations based on the dependent variable, (2) once a rule is selected and a node is split into two, the same process is applied to each 'child' node. (3) splitting stops when CART detects no further gain can be made, or some preset stopping rules are met. Each branch of the tree ends in a terminal node. Each observation falls into exactly one terminal node, and each terminal node is uniquely defined by a set of rules.

Both unsupervised and supervised feature extraction methods are applied to Rotation Forest. Principal Component Analysis (PCA) is the most popular linear unsupervised feature extraction method, which can keep the most information in a few components in terms of variance. Though Cheriyadat and Bruce provide theoretical and experimental analysis to demonstrate that PCA is not optimal for dimensionality reduction in target detection and classification of hyperspectral data, PCA are still competitive for the purpose of classification because of its low complexity and the absence of parameters [16, 24].

Linear Discriminant Analysis(LDA) is the best-known supervised feature extraction approaches. But this method has the limitation: for C class classification problem, it can extract at maximum $C - 1$ features [18, 54]. That means in Rotation Forest, we should define the value of C is greater than K. In order to solve the problem, we adapt Local Fisher Discriminant Analysis (LFDA) instead of LDA. LFDA effectively combines the ideas of LDA and Locality Preserving Projection (LPP), which leads to both maximize between-class separability and preserve with-class local structure [80]. It can be viewed as the following eigenvalue decomposition problem:

$$\mathbf{S}^{lb}\mathbf{v} = \lambda \mathbf{S}^{lw}\mathbf{v} \qquad (6.1)$$

where, $\mathbf{v}$ is an eigenvector and λ is the eigenvalue corresponding to $\mathbf{v}$. $\mathbf{S}_{lb}$ and $\mathbf{S}_{lw}$ denote the local between-class and within-class scatter matrix. LFDA wants to find an eigenvector matrix that maximize the local between-class scatter in the embedding space while minimize the local within-class scatter in the embedding space. $\mathbf{S}_{lb}$ and $\mathbf{S}_{lw}$ can be defined:

$$\mathbf{S}^{lb} = \frac{1}{2} \sum_{i,j=1}^{n} \omega_{i,j}^{lb} (\mathbf{x}_i - \mathbf{x}_j)(\mathbf{x}_i - \mathbf{x}_j)^\top \qquad (6.2)$$

$$\mathbf{S}^{lw} = \frac{1}{2} \sum_{i,j=1}^{n} \omega_{i,j}^{lw} (\mathbf{x}_i - \mathbf{x}_j)(\mathbf{x}_i - \mathbf{x}_j)^\top \qquad (6.3)$$

where, ω^{lb} and ω^{lw} are the weight matrices with:

$$\omega_{i,j}^{lb} = \begin{cases} A_{i,j} \left(\frac{1}{n} - \frac{1}{n_{y_i}} \right) & y_i = y_j \\ \frac{1}{n} & y_i \neq y_j \end{cases} \qquad (6.4)$$

$$\omega_{i,j}^{lw} = \begin{cases} \frac{A_{i,j}}{n_{y_i}} & y_i = y_j \\ 0 & y_i \neq y_j \end{cases} \quad (6.5)$$

where, $A_{i,j}$ is the affinity value between $\mathbf{x}_i$ and $\mathbf{x}_j$ in the local space.

$$A_{i,j} = \exp\left(-\frac{\|\mathbf{x}_i - \mathbf{x}_j\|}{\sigma_i \sigma_j}\right) \quad (6.6)$$

$$\sigma_i = \|\mathbf{x}_i - \mathbf{x}_i^e\| \quad (6.7)$$

where, $\mathbf{x}_i^e$ is the e-th nearest neighbor of $\mathbf{x}_i$, n_{y_i} is the number of labeled samples in class $y_i \in \{1, 2, 3, ..., C\}$.

6.4.2 Rotation SVM

SVM classifier has shown better classification performance for high-dimensional data than other classifier. SVM is very stable that small changes in the training set cannot produce very different SVM classifiers.

Therefore, it is difficult to get an ensemble of multiple SVM that perform better than a single SVM using the state of the art ensemble methods. Thus, we hope to introduce more diversity into SVM. In [52], diversity is analyzed for Random Projections (RP) with and without splitting into group of attributes. Therefore, we introduce Random Projection (RP) into rotation-based SVM in order to promote the diversity within the ensemble.

RP obtains the rotation matrix using simply random number. Unlike other feature extraction methods such as PCA, RP can get a projected space which is bigger than the original. Two types of RP are used in this chapter [1]:

1. Gaussian. Each value in transformation matrix comes from a Gaussian distribution (mean 0 and standard deviation).
2. Sparse. The entry values are $\sqrt{3} \times \alpha$, where, α is a random number taking the following value: -1 with the probability $1/6$, 0 with the probability $2/3$ and $+1$ with probability $1/6$.

6.5 Experimental Results and Analysis

6.5.1 Experimental Setup

In this section, we present the results that we obtained with rotation-based ensemble on different types of images. Two airborne hyperspectral images are used to evaluate

Rotation Forest (RoF). An airborne hyperspectral and a multi-date remote sensing images are applied to test the performance of Rotation SVM (RoSVM). The descriptions of the data are detailed in the following two subsections. Overall accuracy (OA), average accuracy (AA), and class-specific accuracy are used to evaluate the efficiency of RoF and RoSVM.

Popular ensemble methods, including Bagging [9], AdaBoost [28] and Random Forest (RF) [10] are added to be compared with Rotation Forest. The performance achieved by Rotation Forest is illustrated using the following design:

- Number of features in each subset: $M = 10$;
- Number of classifiers in the ensemble: $L = 10$;
- Feature extraction method: PCA [42] and LFDA [80];

we employed RoF-PCA and RoF-LFDA as the abbreviations of Rotation Forest with PCA and LFDA.

Gaussian RBF kernel is adopt in the SVM. In order to reduce the computational time in the ensembles of SVM, we used the fixed parameters in SVM. Random Projection-based ensemble is added to compare with RoSVM using RP projections. Two sizes of projected space dimension have been tested (100 and 150 %). The configurations of 150 % size are denoted as RoSVM or RP 150 %. The performance achieved by RoSVM is illustrated using the following designs:

- Number of features in each subset: $M = 10$;
- Number of classifiers in the ensemble: $L = 10$;
- Feature extraction method: Random Projection (RP) with *Gaussian* and *Sparse*;
- Base classifier: SVM.

In the following experiments, we employed RP and RoSVM as the abbreviations of Random Projection-based ensemble and rotation-based SVM ensemble.

6.5.2 Rotation Forest

6.5.2.1 Indiana Pines AVIRIS Image

The first hyperspectral image is recorded by the Airborne Visible/Infrared Imaging Spectrometer (AVIRIS) sensor over the Indiana Pines in Northwestern Indiana, USA. The image is composed of 145 × 145 pixels, and the spatial resolution is 20 m per pixel. This image is a classical benchmark to validate the accuracy of hyperspectral image analysis algorithms and constitutes a challenging problem due to the significant presence of mixed pixels in all available classes and also because of the unbalanced number of available labeled pixels per class. The three-band color composite and ground truth of AVIRIS image can be seen in Fig. 6.4. We have chosen 20 pixels of each class from the available ground truth (a total size of 320 pixels) as the training set.

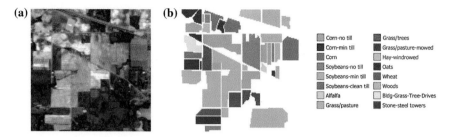

Fig. 6.4 a Three-band color composite of AVIRIS image. **b** Ground truth: Corn-no till, corn-min till, corn, soybean-no till, soybeans-min till, soybeans-clean till, alfalfa, grass/pasture, grass/trees, grass/pasture-mowed, hay-windrowed, oats, wheat, woods, bldg-grass-tree-drives, stone-steel towers

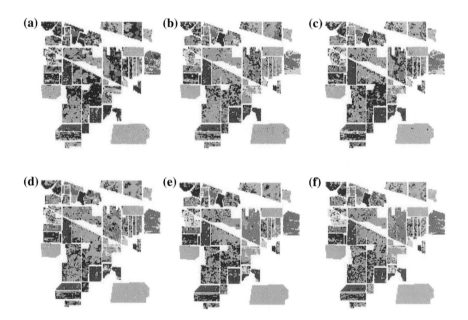

Fig. 6.5 Classification results of Indiana Pines AVIRIS image. Different color represents the different class. The color of the classes can be found in Fig. 6.4. **a** CART. **b** Bagging. **c** AdaBoost. **d** RF. **e** RoF-PCA. **f** RoF-LFDA

Table 6.4 shows the classification accuracies (OA%) obtained by the Rotation Forest approaches as well as other algorithms using different training samples. We highlight the highest accuracies of each case in bold font. From Table 6.4, it can be seen that RoF-PCA and RoF-LFDA achieve better results than other ensemble approaches (Bagging, Adaboost, and RF). Compared to Bagging, AdaBoost and RF, Rotation Forest can promote the diversity and improve the accuracy of individual classifier within the ensemble. Therefore, in most cases, Rotation Forest is superior to Bagging, AdaBoost, and Random Forest. Our experimental results are compati-

Table 6.4 Overall, average and class-specific accuracies of the Indiana Pines AVIRIS image

Class	Train	Test	CART	Bagging	AdaBoost	RF	RoF-PCA	RoF-LFDA
Alfalfa	20	54	77.78	79.63	79.63	87.04	**88.89**	**88.89**
Corn-no till	20	1434	32.15	**52.09**	37.66	37.34	50.28	50.84
Corn-min till	20	834	37.89	37.53	45.92	48.08	**61.51**	54.8
Bldg-grass-tree-drives	20	234	44.44	50.85	44.02	47.44	70.51	**78.21**
Grass/pasture	20	497	46.08	71.23	46.48	63.38	76.46	**78.47**
Grass/trees	20	747	57.43	79.92	**84.47**	73.9	73.63	76.44
Grass/pasture-mowed	20	26	88.46	88.46	92.31	92.31	**96.15**	92.31
Corn	20	489	62.99	49.69	58.69	81.6	**86.09**	84.05
Oats	20	20	30	90	75	**100**	**100**	**100**
Soybeans-no till	20	968	30.48	42.25	49.38	53.71	76.55	**76.65**
Soybeans-min till	20	2468	23.3	35.13	31	**45.79**	31.69	30.88
Soybeans-clean till	20	614	28.66	31.92	34.2	43.81	47.56	**51.79**
Wheat	20	212	86.79	88.68	86.79	92.45	**94.34**	91.98
Woods	20	1294	69.09	76.82	83	83.77	**91.19**	89.49
Hay-windrowed	20	380	48.16	45.79	46.84	**55.53**	48.42	48.42
Stone-steel towers	20	95	74.74	96.84	**97.89**	95.79	93.68	94.74
OA			41.44	51.87	50.54	56.97	**60.88**	60.60
AA			52.4	63.55	63.55	68.87	74.18	**74.25**

ble with the theorectical analysis. For instance, CART, Bagging, Adaboost and RF acquired an OA of 41.44, 51.87, 50.54 and 56.97 %, respectively. RoF-PCA and RoF-LFDA respectively increased the OA to 60.88 and 60.6 %, while the AA of RoF-PCA and RoF-LFDA were improved to 23.78 and 23.85 % percentage points compared to CART. The OA of RoF-PCA is slightly higher than the one of RoF-LFDA. But there is no significantly difference between the two classification results according to McNemar test. Nine of sixteen class-specific accuracies is improved by RoF-PCA and RoF-LFDA.

The classification results of Indiana Pines AVIRIS image are shown in Fig. 6.5. The classification map for the CART classifier was very noisy because CART is not a promising classifier for high-dimensional data. Compared to the reference data presented in Fig. 6.4b, all the ensemble methods produced more corrected classification results than CART. If we carefully look at the reference image, particularly, the area of *Soybean-no till*, this region is almost correctly classified by RoF-PCA and RoF-LFDA, whereas it is classified as *Corn-min till* and *Corn-no till* by other classifiers.

6.5.2.2 Pavia Center DAIS Image

The second image was acquired by the DAIS sensor at 1500 m flight altitude over the city of Pavia, Italy. The image (seen in Fig. 6.6) has a size of 400 × 400 pixels, with ground resolution of 5 m. The 80 data channels recorded by this spectrometer were used for this experiment. Nine land cover classes of interest are considered, which are detailed in Table 6.5, with the number of labeled samples for each class.

Table 6.5 Overall, average and class-specific accuracies of the Pavia Center DAIS image

Class	Train	Test	CART	Bagging	AdaBoost	RF	RoF-PCA	RoF-LFDA
Water	10	4281	98.11	96.15	90.66	**100**	**100**	**100**
Trees	10	2424	56.89	67.12	63.74	88.7	87.83	**91.3**
Meadows	10	1251	97.44	97.76	97.2	**99.52**	99.12	99.12
Bricks	10	2237	74.17	77.08	72.68	65.02	**84.27**	80.2
Soil	10	1475	50.29	60.39	65.38	**76.7**	74.77	74.71
Asphalt	10	1704	77.7	77.7	83.28	77.35	91.99	**97.91**
Bitumen	10	685	68.22	63.66	76.62	83.33	**94.95**	93.74
Parking lot	10	287	70.22	87.01	**91.09**	78.1	86.28	88.32
Shadows	10	241	86.72	66.39	85.06	92.95	**95.85**	89.63
OA			76.71	79.09	79.6	87.67	91.72	**91.8**
AA			75.53	77.03	80.63	84.63	**90.56**	90.55

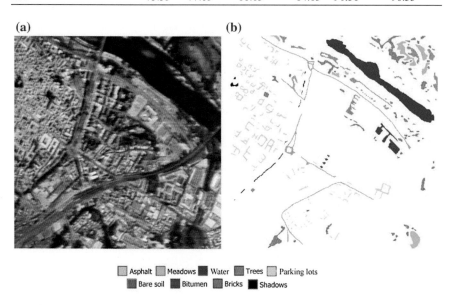

Fig. 6.6 **a** Three-band color composite of DAIS image. **b** Ground truth

The global accuracies and class-specific accuracies of the Pavia Center DAIS image are reported in Table 6.5. The classification results achieved by the ensemble classifiers are similar with the ones of AVIRIS image. Regarding the overall accuracies, Rotation Forest with different feature extraction algorithms are all superior to other approaches under comparison. RoF-LFDA yields the highest OA (91.8%). The accuracies of class *Bricks*, *Asphalt* and *Bitumen* are significantly improved by the Rotation Forest ensemble classifiers. The classification results of Pavia Center DAIS image are shown in Fig. 6.7.

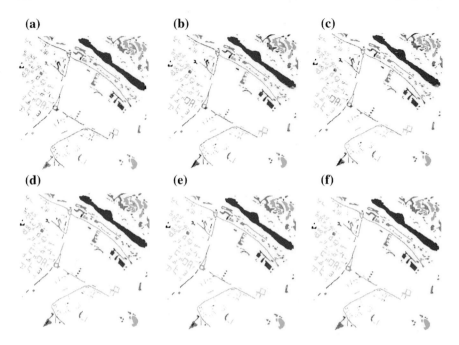

Fig. 6.7 Classification results of Pavia Center DAIS image. **a** CART. **b** Bagging. **c** AdaBoost. **d** RF. **e** RoF-PCA. **f** RoF-LFDA

6.5.2.3 Sensitivity of the Parameters

Ensemble size (T) and the number of features in a subset (M) are the key parameters of Rotation Forest, which are indicators of the operating complexity. In order to investigate the impacts of these parameters, we have performed the classification results using different ensemble size when the number of subset M is fixed to 10, different number of features in a subset when ensemble size T is fixed to 10. Fig. 6.8 shows the OA (%) using different number of T and M obtained from AVIRIS and DAIS images. For AVIRIS image, the classification results are improved when the values of T and M increased. RoF-PCA is superior to RoF-LFDA. For DAIS image, the OAs are improved with the increasement of T. The general trend of different values of M is not obvious.

6.5.2.4 Discussion

Based on the above classification results, we identify that the incorporation of multiple classifiers has shown great improvement for the classification of high-dimensional data. In order to make MCS effective, we should enforce the diversity by the manipulation of training sets. Bagging and Boosting aim at changing the distribution of the training samples to obtain the different training set. Random subspace method

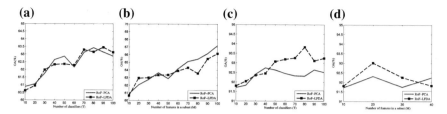

Fig. 6.8 Effects of varying parameters on the performance of rotation forest. Indiana Pines AVIRIS image. **a** Sensitivity for change of T ($M = 10$). **b** Sensitivity for change of M ($T = 10$). Pavia Center DAIS image. **c** Sensitivity for change of T ($M = 10$). **d** Sensitivity for change of M ($T = 10$)

constructs several classifier by random selecting the subset of the features. It is very useful for the classification problem where the number of features is much larger than the number of training samples. Random subspace method is a generalization of the Random Forest algorithm, whereas Random Forest is composed of decision trees. Rotation Forest tries to create the individual classifiers that are both diverse and accurate, each based on a different axis rotation of attributes. To create different training set, the features are randomly split into a given number of subsets and feature extraction is applied to each subset. Decision trees is very sensitive to the rotation of axis. In this chapter, we select CART to construct Rotation Forest. Rotation Forest can promote more diversity than Bagging, AdaBoost and Random Forest. Therefore, it can produce more accurate results than Bagging, AdaBoost and Random Forest. An important issue of Rotation Forest is the selection of the parameters (T and M). A larger value of T will often increase the accuracy and also increase the computation time. The optimal value of M is hard to determine. Different datasets achieve the highest accuracy with different value of M. The computation time of Rotation Forest approaches is longer than those of Bagging, AdaBoost and Random Forest. But the computation complexity of Rotation Forest is much less than the one of the strong classifier of high-dimensional data, such as SVM.

6.5.3 Rotation SVM

6.5.3.1 Indiana Pines AVIRIS Image

Table 6.6 shows overall, average and class-specific accuracies using different version of rotation-based SVMs. We highlight the highest accuracies of each case in bold font. It can be seen that RoSVM achieve the better results than RP, because RoSVM can provide more diversity than RP. For this dataset, RP with Gaussian is superior to the one of Sparse. By employing a slightly higher size of a projected space, the results of RP is improved but RoSVM yields bad results. The corresponding results are shown in Fig. 6.9. We have studied the impacts of T and M in RoSVM. The sensitivity of performance using different T and M is not obvious.

Table 6.6 Overall, average and class-specific accuracies of the Indiana Pines AVIRIS image

Class	SVM	RP				RoSVM			
		Gaussian	Gaussian 150 %	Sparse	Sparse 150 %	Gaussian	Gaussian 150 %	Sparse	Sparse 150 %
Alfalfa	85.19	90.74	88.89	94.44	88.88	100	94.44	94.44	88.89
Corn-no till	23.78	62.41	**66.46**	57.32	53.84	57.74	56	51.88	62.76
Corn-min till	48.44	52.39	57.19	51.68	45.8	57.55	**62.23**	56.36	55.88
Bldg-Grass-Tree-Drives	78.63	80.77	79.06	73.50	65.81	80.77	70.51	**84.19**	64.53
Grass/pasture	79.07	78.07	81.09	84.51	77.87	88.33	87.32	88.13	**89.74**
Grass/trees	77.24	77.78	85.14	88.09	88.22	91.97	89.56	92.24	**93.17**
Grass/pasture-mowed	92.31	100	100	96.15	100	100	96.15	100	96.15
Corn	85.48	68.30	90.39	82	74.44	95.09	92.64	92.23	**96.11**
Oats	75	100	100	95	95	100	100	100	90
Soybeans-no till	45.35	57.85	58.99	50.20	55.37	**65.08**	60.54	60.12	60.02
Soybeans-min till	52.92	45.83	36.47	40.44	**53.53**	49.31	52.15	48.87	46.31
Soybeans-clean till	39.74	57.98	**64.01**	62.38	48.53	59.45	47.23	58.14	41.86
Wheat	98.58	98.11	99.06	95.75	98.11	98.58	97.17	99.53	**99.06**
Woods	92.74	84.18	**93.12**	79.6	75.35	90.57	83.93	92.35	87.86
Hay-windrowed	28.16	57.89	51.58	64.73	**71.57**	57.63	63.94	65.53	59.21
Stone-steel towers	96.84	92.63	87.37	93.68	94.74	96.84	100	96.84	88.42
OA	57.87	63.39	65.1	62.07	62.81	**68.42**	66.87	66.85	66.15
AA	68.72	75.31	77.43	75.59	74.19	**80.56**	78.36	80.05	76.25

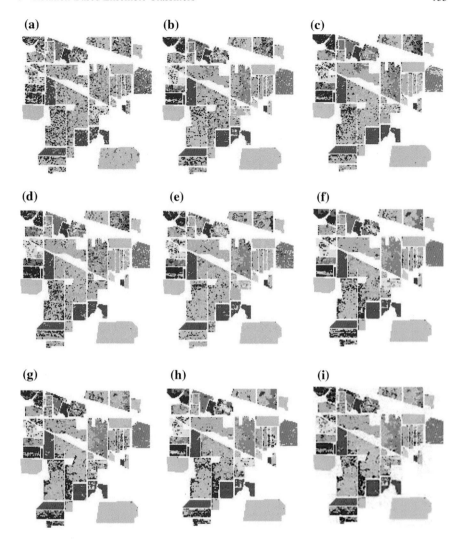

Fig. 6.9 Classification results of Indiana Pines AVIRIS image. **a** SVM. **b** RP Gaussian. **c** RP Gaussian 150 %. **d** RP Sparse. **e** RP Sparse 150 %. **f** RoSVM Gaussian. **g** RoSVM Gaussian 150 %. **h** RoSVM Sparse. **i** RoSVM Sparse 150 %

6.5.3.2 CHRIS Multi-date Images

The second high-dimensional data is the three dates of Compact High-Resolution Imaging Spectrometer (CHRIS) images acquired by the Project for On-Board Autonomy (PROBA)-1 satellite with spatial resolution of 21 m/pixel. The total number of spectral bands is 54. More details about CHRIS image can be seen in [23]. Training samples contains 2,297 samples and test data includes 1,975 samples with 11 classes.

The flowchart of RoSVM for classifying CHRIS image is the same as the previous AVIRIS dataset. Single SVM achieved the accuracy of 84.05 %. All the methods based on RP and RoSVM ensemble can generate the better accuracies than a single SVM. In particular, RoSVM ensembles are slightly superior to RP ensembles because they enforce the diversity by applying RP to the subsets of the features. RoSVM with Spare RP gains the highest the overall accuracy.

6.5.3.3 Discussion

SVM is a stable classifier, so it is hard to generate different individual SVM classifiers using the common manipulation ways. Therefore, we should introduce more diversity to construct the diverse individual SVM classifiers. In this chapter, we adapt Random Projection methods to produce diverse SVM classifiers. Two sizes of projected space dimension have been tested. Experimental results indicated that RoSVM ensemble outperform RP ensembles. The main drawback of RoSVM is the computational complexity, especially for large training samples. The sensitivity of performance using different M is not obvious.

6.6 Conclusion

In this chapter, we first presented a review of MCS approaches with special focus on applications of high-dimensional data. Recently rotation-based ensemble classifiers were applied to high-dimensional data. They consist in splitting the feature set into several subsets, running feature extraction algorithms separately on each subset and then reassembling a new extracted feature set while keeping all the components. CART Decision Tree and SVM classifiers are used as the base classifier. Different splits of the feature set lead to different rotations. Thus diverse classifiers are obtained. We take into account both diversity and accuracy. Rotation Forest using PCA, LFDA, Rotation SVM using RP are used to classify high-dimensional data.

Experimental results have shown that rotation-based ensemble methods (both DT and SVM) outperform classical ensemble methods such as Bagging, AdaBoost, Random Forest in terms of accuracies. The key parameters are also explored in this chapter. Future studies will be devoted to the integration of rotation-based ensemble classifiers with other ensemble approaches, the selection of an optimized Decision Tree model, and the use of other effective feature extraction algorithms.

References

1. Achlioptas, D (2001) Database-friendly random projections. In: Proceedings of the twentieth ACM SIGMOD-SIGACT-SIGART symposium on principles of database systems, New York, NY, USA, 2001, pp 274–281
2. Aksela M, Laaksonen J (2006) Using diversity of errors for selecting members of a committee classifier. Pattern Recogn 39(4):608–623

3. Bakos KL, Gamba P (2011) Hierarchical hybrid decision tree fusion of multiple hyperspectral data processing chains. IEEE Trans Geosci Remote Sens 49(1–2):388–394
4. Benediktsson JA, Chanussot J, Fauvel M (2007) Multiple classifier systems in remote sensing: from basics to recent developments. Lect Notes Comput Sci 4472:501–512
5. Benediktsson JA, Sveinsson JR (1997) Hybrid consensus theoretic classification. IEEE Trans Geosci Remote Sens 35(4):833–843
6. Benediktsson JA, Swain PH (1992) Consensus theoretic classification methods. IEEE Trans Syst Man Cybern 22(4):688–704
7. Biggio B, Fumera G, Roli F (2010) Multiple classifier systems for robust classifier design in adversarial environments. J Mach Learn Cybern 1:27–41
8. Braun AC, Weidner U, Hinz S (2012) Classification in high-dimensional feature spaces assessment using SVM, IVM and RVM with focus on simulated EnMap data. IEEE J Sel Top Appl Earth Observations Remote Sens 5(2):436–443
9. Breiman L (1996) Bagging predictors. Mach Learn 24(2):123–140
10. Breiman L (2001) Random forest. Mach Learn 45(1):5–32
11. Briem G, Benediktsson J, Sveinsson J (2002) Multiple classifiers applied to multisource remote sensing data. IEEE Trans Geosci Remote Sens 40(10):2291–2299
12. Brown G, Wyatt J, Harris R, Yao X (2005) Diversity creation methods: a survey and categorisation. Inf Fusion 6(1):5–20
13. Bryll RK, Gutierrez-Osuna R, Quek FKH (2003) Attribute bagging: improving accuracy of classifier ensembles by using random feature subsets. Pattern Recogn 36(6):1291–1302
14. Chan JC, Paelinckx D (2008) Evaluation of Random Forest and AdBboost tree-based ensemble classification and spectral band selection for ecotope mapping using airborne hyperspectral imagery. Remote Sens Environ 112(6):2999–3011
15. Chandra A, Yao X (2006) Evolving hybrid ensembles of learning machines for better generalisation. Neurocomputing 69(7–9):686–700
16. Cheriyadat A, Bruce LM (2003) Why principal component analysis is not an appropriate feature extraction method for hyperspectral data. In: Proceedings of IEEE geoscience and remote sensing symposium (IGARSS), Toulouse, France, 2003, pp 3420–3422
17. Cunningham P, Carney J (2000) Diversity versus quality in classification ensembles based on feature selection. In: 11th European conference on machine learning, Barcelona, Spain, 2000, pp 109–116
18. Dell'Acqua F, Gamba P, Ferari A, Palmason BJA, Arnason K (2004) Exploiting spectral and spatial information in hyperspectral urban data with high resolution. IEEE Geosci Remote Sens Lett 1(4):322–326
19. Dietterich TG, Bakiri G (1995) Solving multiclass learning problems via error-correcting output codes. J Artif Intell Res 2(1):263–286
20. Doan HTX, Foody GM (2007) Increasing soft classification accuracy through the use of an ensemble of classifiers. Int J Remote Sens 28(20):4609–4623
21. Du P, Xia J, Zhang W, Tan K, Liu Y, Liu S (2012) Multiple classifier system for remote sensing image classification: a review. Sensors 12(4):4764–4792
22. Du P, Zhang W, Xia J (2011) Hyperspectral remote sensing image classification based on decision level fusion. Chin Optics Lett 9(3):031002–031004
23. Duca R, Frate FD (2008) Hyperspectral and multiangle CHRIS-PROBA images for the generation of land cover maps. IEEE Trans Geosci Remote Sens 10:2857–2866
24. Fauvel M, Chanussot J, Benediktsson JA (2009) Kernel principal component analysis for the classification of hyperspectral remote sensing data over urban areas. EURASIP J Adv Signal Process 2:1–15
25. Fleiss J (1981) Statistical methods for rates and proportions. Wiley, New York
26. Foody GM (2009) Classification accuracy comparison: hypothesis tests and the use of confidence intervals in evaluations of difference, equivalence and non-inferiority. Remote Sens Environ 113(8):1658–1663
27. Foody GM, Boyd DS, Sanchez-Hernandez C (2007) Mapping a specific class with an ensemble of classifiers. Int J Remote Sens 28:1733–1746

28. Freund Y, Schapire RE (1996) Experiments with a new Boosting algorithm. In: International conference on machine learning, Bari, Italy, 1996, pp 148–156
29. Gao J, Liang F, Fan W, Sun Y, Han J (2013) A graph-based consensus maximization approach for combining multiple supervised and unsupervised models. IEEE Trans Knowl Data Eng 25(1):15–28
30. Giacinto G, Roli F (2001) Design of effective neural network ensembles for image classification. Image Vis Comput J 19(9/10):697–705
31. Giacinto G, Roli F, Fumera G (2000) Selection of image classifiers. Electron Lett 36:420–422
32. Giacinto G, Roli F, Vernazza G (1997) Comparison and combination of statistical and neural network algorithms for remote-sensing image classification. Neurocomputation in remote sensing data analysis. Springer, Berlin, pp 117–124
33. Gislason PO, Benediktsson JA, Sveinsson JR (2006) Random forests for land cover classification. Pattern Recogn Lett 27(4):294–300
34. Gmez-Chova L, Fernndez-Prieto D, Calpe-Maravilla J, Soria-Olivas E, Vila-Francs J, Camps-Valls G (2006) Urban monitoring using multi-temporal SAR and multi-spectral data. Pattern Recogn Lett 27(4):234–243
35. Guyon I, Gunn S, Nikravesh M, Zadeh L (2006) Feature extraction: foundations and applications. Springer, New York
36. Ham J, Chen Y, Crawford MM, Ghosh J (2005) Investigation of the random forest framework for classification of hyperspectral data. IEEE Trans Geosci Remote Sens 43(3):492–501
37. Hansen LK, Salamon P (1990) Neural network ensembles. IEEE Trans Pattern Anal Mach Intell 12(10):993–1001
38. He Z, Xu X, Deng S (2005) A cluster ensemble method for clustering categorical data. Inf Fusion 6(2):143–151
39. Ho TK (1998) The random subspace method for constructing decision forests. IEEE Trans Pattern Anal Mach Intell 20(8):832–844
40. Hughes G (1968) On the mean accuracy of statistical pattern recognizers. IEEE Trans Inf Theory 14(1):55–63
41. Iam-On N, Boongoen T, Garrett S, Price C (2011) A link-based approach to the cluster ensemble problem. IEEE Trans Pattern Anal Mach Intell 33(12):2396–2409
42. Richards JA, Jia X (2006) Remote sensing digital image analysis: an introduction. Springer, New York
43. Kang H-J, Lee S-W (2000) An information-theoretic strategy for constructing multiple classifier systems. In: International conference on pattern recognition (ICPR), Barcelona, Spain, 2000, pp 2483–2486
44. Kawaguchi K, Nishii R (2007) Hyperspectral image classification by bootstrap AdaBoost with random decision stumps. IEEE Trans Geosci Remote Sens 45(11–2):3845–3851
45. Kohavi R, Wolpert DH (1996) Bias plus variance decomposition for zero-one loss functions. In: 13th international conference on machine learning (ICML), Bari, Italy, 1996, pp 275–283
46. Kong Z, Cai Z (2007) Advances of research in fuzzy integral for classifiers' fusion. In: Proceedings of the eigth ACIS international conference on software engineering, artificial intelligence, networking, and parallel/distributed computing, Washington, DC, USA pp 809–814
47. Krogh A, Vedelsby J (1995) Neural network ensembles, cross validation, and active learning. Adv Neural Inf Process Syst 7:231–238
48. Kumar S, Ghosh J, Crawford MM (2002) Hierarchical fusion of multiple classifiers for hyperspectral data analysis. Pattern Anal Appl 5:210–220
49. Kuncheva L, Whitaker C, Shipp C, Duin R (2000) Is independence good for combining classifiers? In: International conference on pattern recognition (ICPR), 2000, p 2168
50. Kuncheva LI (2004) Combining pattern classifiers: methods and algorithms. Wiley-Interscience, New Jersey
51. Kuncheva LI (2005) Diversity in multiple classifier systems. Inf Fusion 6(1):3–4
52. Kuncheva LI, Rodriguez JJ (2007) An experimental study on rotation forest ensembles. In: Proceedings of the 7th international workshop on multiple classifier systems, Prague, Czech Republic, 2007, pp 459–468

53. Kuncheva LI, Whitaker CJ (2003) Measures of diversity in classifier ensembles and their relationship with the ensemble accuracy. Mach Learn 51(2):181–207
54. Landgrebe DA (1984) Signal theory methods in multispectral remote sensing. Wiley, New York
55. Lawrence R, Bunna A, Powellb S, Zambon M (2004) Classification of remotely sensed imagery using stochastic gradient boosting as a refinement of classification tree analysis. Remote Sens Environ 90(3):331–336
56. Lu D, Weng Q (2007) A survey of image classification methods and techniques for improving classification performance. Int J Remote Sens 28(5):823–870
57. Margineantu DD, Dietterich TG (1997) Pruning adaptive boosting. In: Proceedings of the fourteenth international conference on machine learning, San Francisco, CA, USA, 1997, pp 211–218
58. Martinez-Munoz G, Suarez A (2005) Switching class labels to generate classification ensembles. Pattern Recogn 38(10):1483–1494
59. Melgani F, Bruzzone L (2004) Classification of hyperspectral remote sensing images with support vector machines. IEEE Trans Geosci Remote Sens 42(8):1778–1790
60. Michail P, Benediktsson JA, Ioannis K (2002) The effect of classifier agreement on the accuracy of the combined classifier in decision level fusion. IEEE Trans Geosci Remote Sens 39(11):2539–2546
61. Mojaradi B, Abrishami-Moghaddam H, Zoej M, Duin R (2009) Dimensionality reduction of hyperspectral data via spectral feature extraction. IEEE Trans Geosci Remote Sens 47(7):2091–2105
62. Moon H, Ahn H, Kodell RL, Baek S, Lin C-J, Chen JJ (2007) Ensemble methods for classification of patients for personalized medicine with high-dimensional data. Artif Intell Med 41(3):197–207
63. Moreno-Seco F, Inesta J, Ponce De Leon P, Mico L (2006) Comparison of classifier fusion methods for classification in pattern recognition tasks. Lect Notes Comput Sci 4109:705–713
64. Navalgund RR, Jayaraman V, Roy PS (2007) Remote sensing applications: an overview. Current 93(12):1747–1766
65. Nemmour H, Chibani Y (2006) Multiple support vector machines for land cover change detection: an application for mapping urban extensions. ISPRS J Photogrammetry Remote Sens 61(2):125–133
66. Ozcift A, Gulten A (2011) Classifier ensemble construction with rotation forest to improve medical diagnosis performance of machine learning algorithms. Comput Methods Programs Biomed 104(3):443–451
67. Pal M, Foody GM (2010) Feature selection for classification of hyperspectral data by SVM. IEEE Trans Geosci Remote Sens 48(5):2297–2307
68. Partridge D, Krzanowski W (1997) Software diversity: practical statistics for its measurement and exploitation. Inf Softw Technol 39(10):707–717
69. Plaza A, Martinez P, Plaza J, Perez R (2005) Dimensionality reduction and classification of hyperspectral image data using sequences of extended morphological transformations. IEEE Trans Geosci Remote Sens 43(3):466–479
70. Rahman A, Fairhurst M (1999) Serial combination of multiple experts: a unified evaluation. Pattern Anal Appl 2:292–311
71. Ranawana R, Palade V (2006) Multi-classifier systems: review and a roadmap for developers. Int J Hybrid Intell Syst 3(1):1–41
72. Rodriguez JJ, Kuncheva LI (2009) Rotation forest: a new classifier ensemble method. IEEE Trans Pattern Anal Mach Intell 28(10):1619–1630
73. Rodriguez-Galianoa VF, Ghimireb B, Roganb J, Chica-Olmoa M, Rigol-Sanchezc JP (2011) An assessment of the effectiveness of a random forest classifier for land-cover classification. ISPRS J Photogrammetry Remote Sens 67:93–104
74. Rokach L (2010) Pattern classification using ensemble methods. World Scientific, Singapore
75. Ruitenbeek F, Debba P, Meer F, Cudahy T, Meijde M, Hale M (2006) Mapping white micas and their absorption wavelengths using hyperspectral band ratios. Remote Sens Environ 102(3–4):211–222

76. Ruta D, Gabrys B (2000) An overview of classifier fusion methods. Comput Inf Syst 7(1):1–10
77. Schölkopf B, Platt JC, Shawe-Taylor J, Smola AJ, Williamson RC (2001) Estimating the support of a high-dimensional distribution. Neural Comput 13(7):1443–1471
78. Smits PC (2002) Multiple classifier systems for supervised remote sensing image classification based on dynamic classifier selection. IEEE Trans Geosci Remote Sens 40:801–813
79. Steele BM (2000) Combining multiple classifiers: an application using spatial and remotely sensed information for land cover type mapping. Remote Sens Environ 74(3):545–556
80. Sugiyama M (2007) Dimensionality reduction of multimodal labeled data by local fisher discriminant analysis. J Mach Learn Res 27(8):1021–1064
81. Sun Q, Ye XQ, Gu WK (2000) A new combination rules of evidence theory (in chinese). Acta Electronica Sinica 8(1):117–119
82. Tsoumakas G, Partalas I, Vlahavas I (2009) An ensemble pruning primer. In: Okun O, Valentini G (eds) Applications of supervised and unsupervised ensemble methods. Springer, Berlin, pp 1–13
83. Vapnik VN (1995) The nature of statistical learning theory. Springer, Berlin
84. Waske B, Benediktsson JA, Arnason K, Sveinsson JR (2009) Mapping of hyperspectral aviris data using machine-learning algorithms. Can J Remote Sens 35:106–116
85. Waske B, Van Der Linden S, Benediktsson JA, Rabe A, Hostert P (2010) Sensitivity of support vector machines to random feature selection in classification of hyperspectral data. IEEE Trans Geosci Remote Sens 48(7):2880–2889
86. Wolpert DH, Macready WG (1997) No free lunch theorems for optimization. IEEE Trans Evol Comput 1(1):67–82
87. Woods K, Kegelmeyer WP, Bowyer K (1997) Combination of multiple classifiers using local accuracy estimates. IEEE Trans Pattern Anal Mach Intell 19(4):405–410
88. Wozniak M, Grana M, Corchado E (2014) A survey of multiple classifier systems as hybrid systems. Inf Fusion 16(1):3–17
89. Xia J, Du P, He X, Chanussot J (2014) Hyperspectral remote sensing image classification based on rotation forest. IEEE Geosci Remote Sens Lett 11:239–243
90. Xu L, Krzyzak A, Suen CY (1992) Methods of combining multiple classifiers and their applications to handwriting recognition. IEEE Trans Syst Man Cybern 22(3):418–435
91. Yang J-M, Kuo B-C, Yu P-T, Chuang C-H (2010) A dynamic subspace method for hyperspectral image classification. IEEE Trans Geosci Remote Sens 48(7):2840–2853
92. Zeng H, Triosell HJ (2004) Dimensionality reduction in hyperspectral image classification. In: IEEE International Conference on Image Processing (ICIP), Singapore, 2004, pp 913–916
93. Zhang Y (2010) Ten years of technology advancement in remote sensing and the research in the crc-agip lab in gce. Geomatica 64(2):173–189

Chapter 7
Multimodal Fusion in Surveillance Applications

Virginia Fernandez Arguedas, Qianni Zhang and Ebroul Izquierdo

Abstract The recent outbreak of vandalism, accidents and criminal activities has increased general public's awareness about safety and security, demanding improved security measures. Smart surveillance video systems have become an ubiquitous platform which monitors private and public environments, ensuring citizens well-being. Their universal deployment integrates diverse media and acquisition systems, generating daily an enormous amount of multimodal data. Nowadays, numerous surveillance applications exploit multiple types of data and features benefitting from their uncorrelated contributions. Hence, the analysis, standardisation and fusion of complex content, specially visual, have become a fundamental problem to enhance surveillance systems by increasing their accuracy, robustness and reliability. During this chapter, an exhaustive survey of the existing multimodal fusion techniques and their applications in surveillance is provided. Addressing some of the revealed challenges from the state of the art, this chapter focuses on the development of a multimodal fusion technique for automatic surveillance object classification. The proposed fusion technique exploits the benefits of a Bayesian inference scheme to enhance surveillance systems' performance. The chapter ends with an evaluation of the proposed Bayesian-based multimodal object classifier against two state-of-the-art object classifiers to demonstrate the benefits of multimodal fusion in surveillance applications.

V. Fernandez Arguedas (✉) · Q. Zhang · E. Izquierdo
Multimedia and Vision Research Group, School of Electronic Engineering
and Computer Science, Queen Mary, University of London, Mile End Road, London E1 4NS, UK
e-mail: virginia.fernandez@eecs.qmul.ac.uk; virginia.fernandez-arguedas@jrc.ec.europa.eu

V. Fernandez Arguedas
European Commission—Joint Research Centre (JRC), Via E. Fermi, 21027 Ispra, VA, Italy

Q. Zhang
e-mail: qianni.zhang@eecs.qmul.ac.uk

E. Izquierdo
e-mail: ebroul.izquierdo@eecs.qmul.ac.uk

7.1 Introduction

Recently, society awareness of citizens' security has undergone an exponential increase due to the recent outbreak of vandalism, accidents and criminal activities. The need for enhanced security and the ever-increasing efforts to palliate the threat triggered the deployment of surveillance systems. Such systems are envisaged as surveillance networks, where numerous plugged-in devices are deployed to monitor urban areas and detect suspicious events. Hence, in-growing telecommunications infrastructures are contemplated to ensure the citizens' security and raise society's well-being, leaving behind single camera systems and their limited capabilities. During the years, the type of devices used to monitor urban areas has evolved towards more technological applications, from police records and citizens testimonials to speed sensors and surveillance cameras, creating a colourful spectrum (refer to Fig. 7.1). Additionally, some devices intricate a huge variability between models, e.g., vision cameras and infrared cameras, resulting in a broad amalgam of data and patterns extracted with different standards and presenting complementary information.

The creation of a multimodal surveillance network, fed by multiple inputs, is an open-research challenge which envisages the improvement of the existing surveillance systems as well as strengthening the existing forensic applications. This is not only a benefit of adding new modalities of information but also potentiates the synergy between the complementary information provided by different sensors' types. Nowadays, diverse examples of surveillance networks are deployed all over the world, from distributed networks to centralised architectures, and presenting different topologies [13]. In the literature, several authors have addressed the creation of surveillance multimodal networks, structuring the surveillance process into three steps: (i) extracting useful metadata from videos, (ii) fusing the extracted information from multiple sensors and (iii) presenting them in a user-friendly manner [20]. In [34], authors propose to combine audio and video sensors to enhance the surveillance system robustness by complementing their information in case of sparse camera networks, addressing cost-reduction in surveillance networks by camera control based on audio event detection. Another multimodal surveillance network was proposed by Drajic and Cvejic [14]. Their network, built over visible surveillance and infrared cameras, addressed image fusion to extend the complementary information present in different visual sensors, proposing a region-based image fusion algorithm based on Dual-Tree Complex Wavelet transform. Prati et al. proposed a multimodal sensor network integrating passive infrared sensors with traditional surveillance video cameras for the monitoring of specific situations such as door access and moving object or motion direction changes during occlusions [38]. Their work reported a drastic improvement in accuracy and robustness due to the fusion of complementary information in situations when the visual analysis is challenging.

Despite the fact that the exploitation of complementary sensors attracts a lot of attention both from industrial and academic research community due to its relevant potentiality, smart surveillance systems are mainly built over surveillance cameras,

Fig. 7.1 Surveillance systems multimodal inputs

relying on computer vision techniques for monitoring urban areas. The relative low cost of the employed devices comes together with certain limitations affecting the visual processing components, such as light variance, low resolution, occlusions, low quality, etc. In fact, surveillance cameras, typically installed on fixed and mobile devices, provide a huge quantity of information that has to be contrasted, correlated and integrated in order to react to special situations. Nowadays single camera scenarios are studied to address visual challenges, such as occlusions or low quality, however, there is a strong interest on high-complexity scenarios composed of several visual sensors. Multicamera surveillance networks are increasingly present in the literature addressing surveillance tasks such as scene monitoring, event detection or object tracking [1]. More importantly, they deal with some specific challenges such as overlapped field of views or blank-spaces in adjacent cameras and visual features variability amongst cameras (which are hindering the identification and recognition of moving objects in non-overlapped cameras).

The exponential increase of surveillance data generated everyday by an in-growing set of multimodal sensors is a bottle-neck problem, where the collected information cannot be processed in real-time or even for the archives. An additional problem is the selection of the appropriate information fusion technique required to integrate the sensors plugged-in a surveillance system. The diversity and variability of the surveillance systems deployed require robust but adaptive information fusion techniques which are capable of combining complementary information in versatile environments. In this chapter, the foundation for a Multimodal Surveillance System (MSS) is proposed, stating the requirements, characteristics and identifying the challenges, presenting the potentials in surveillance networks. Furthermore, the crucial need to combine multimodal information to exploit the synergy amongst acquisition systems is addressed in the proposed Bayesian-based Multimodal Fusion technique. Surveillance systems are by nature adaptable and scalable networks which should be able to evolve on time. Hence, an information fusion technique capable of addressing scalability, adaptability, robustness and information incompleteness is presented. Finally, the overall functionality, implementation and evaluation presented in this chapter are framed on a surveillance use case, i.e., object classification in urban scenarios. Its relevance in real-time and forensic applications as well as its impact on event detection algorithms locates it as a foundational step. The exponential increase of surveillance data generated everyday by an in-growing set of multimodal sensors is a bottle-neck problem, where the collected information cannot be processed in real-time or even for the archives. An additional problem is the selection of the appropriate information fusion technique required to integrate the sensors plugged-in a surveillance system. The diversity and variability of the surveillance systems deployed require robust but adaptive information fusion techniques which are capable of combining complementary information in versatile environments. In this chapter, the foundation for a MSS is proposed, stating the requirements, characteristics and identifying the challenges, presenting the potentials in surveillance networks. Furthermore, the crucial need to combine multimodal information to exploit the synergy amongst acquisition systems are addressed in the proposed Bayesian-based Multimodal Fusion technique. Surveillance systems are by nature adaptable and scalable networks which should be able to evolve on time. Hence, an information fusion technique capable of addressing scalability, adaptability, robustness and information incompleteness is presented. Finally, the overall functionality, implementation and evaluation presented in this chapter are framed on a surveillance use case, i.e., object classification in urban scenarios. Its relevance in real-time and forensic applications as well as its impact on event detection algorithms locates it as a foundational step. The exponential increase of surveillance data generated everyday by an in-growing set of multimodal sensors is a bottle-neck problem, where the collected information cannot to be processed in real-time or even for the archives. An additional problem is the information fusion technique required to integrate the sensors plugged-in a surveillance system. The diversity and variability of the surveillance systems deployed require robust but adaptive information fusion techniques which are capable of combining complementary information in versatile environments. In this chapter, the foundation for a MSS is proposed, stating the requirements, characteristics and identifying the challenges, presenting the potentials in surveillance networks. Furthermore, the crucial need to combine multimodal information to exploit the synergy amongst acquisition systems is addressed in the proposed Bayesian-based Multimodal Fusion technique. Surveillance systems are by nature adaptable and scalable networks which should be able to evolve on time. Hence, an information fusion technique capable of addressing scalability, adaptability, robustness and information incompleteness is presented. Finally, the overall functionality, implementation and evaluation presented in this chapter are framed on a surveillance use case, object classification in urban scenarios. Its relevance in real-time and forensic applications as well as its impact on event detection algorithms locates it as a foundational step.

The remainder of the book chapter is organised as follows. An exhaustive survey of the existing multimodal fusion techniques with special attention to object classification and other surveillance applications is presented in Sect. 7.2. The proposed MSS is presented in Sect. 7.3, where the inputs of the proposed surveillance network and the proposed Bayesian-based Multimodal Fusion technique are further detailed. A comprehensive description of the experiments conducted to evaluate the proposed object classifiers is presented in Sect. 7.4. Whilst, Sect. 7.5 draws conclusions and presents the potential future work.

7.2 Multimodal Fusion Techniques in Surveillance Applications

Multimedia analysis and more specifically surveillance systems benefit from different inputs. Such inputs can be captured by different media and present different types of information. The variety in the available media and the different nature of information have motivated multimodal fusion research. Over the past several decades, many different approaches have been proposed to automatically represent objects or concepts in videos, such as visual appearance, motion, shape or temporal evolution [18]. Single features or inputs are capable of obtaining high accuracy results and tackle specific problems, e.g., object detection. However, the use of complementary information enhances the possibilities and capabilities of different systems to perform more sophisticated tasks, e.g., object classification, speaker identification, etc, and increases the accuracy of the overall decision-making process, motivating the research on multimedia fusion.

The variety of media, features or partial decisions provide a wide range of options to address specific tasks. However, the different characteristics of the involved modalities hinder the combination for several reasons including (i) the particular format acquisition of different media, (ii) the confidence level associated to each data depending on the task under analysis, (iii) the independent protection of each type of data and (iv) the different processing times related to the different type of media streams. Due to the existing challenges in multimodal fusion and the range of application tasks, the multimodal fusion techniques in the literature can be categorised according to the *level of fusion (early fusion)* or the *nature of the methods (late fusion)*. On one hand, the former category, *level of the fusion (early fusion)*, includes all the approaches which combine the available input data before performing the objective task. In this case, the number of features extracted from different modalities must be combined in a unique vector (output) which will be considered as a unique input by the objective task. Amongst the advantages of the feature level multimodal fusion techniques, the need for a unique learning phase on the combined feature vector and the possibility to take advantage of the correlation between multiple features from different modalities excel [42]. Despite the advantages, feature level multimodal fusion presents several disadvantages including: (i) the difficulty to learn cross-correlation amongst features increases with the number of different media considered, (ii) before combining features, their format should be the same and (iii) the synchronisation between features is more complex due to their different modalities and non-linearity [47]. On the other hand, the latter category, *nature of the methods (late fusion)*, proposes to analyse each input individually, providing local decisions. Those decisions are then combined using a decision fusion unit to make a fused decision vector that is analysed to obtain a final decision, considering it as the output of the fusion technique. Unlike feature level fusion techniques, decision-level multimodal fusion techniques benefit from unique representations despite the multiple media modalities easing their fusion, the scalability of the system and enabling the use of different and most suitable techniques to obtain partial solutions. However, the acquisition of partial solutions prevents consideration of the features correlation and is affected by the individual learning process associated to each feature. In order to exploit the advantages of both fusion levels, hybrid systems have been proposed. For further information on the state of the art refer to [5, 12, 32, 41].

Considering the exponential growth of the types and amount of media, smart surveillance systems try to convey information captured by different means to achieve a higher robustness and accuracy. In the proposed use case, object classification in urban environments, different acquisition systems are distributed, such as speed-detection sensors, acoustic sensors, videocameras, etc, collecting relevant information. Several classification-based multimodal fusion methods have been proposed in the literature in an attempt to categorise the multimodal input data into one of the pre-defined classes associated to the application under analysis. However, the most popular multimodal fusion techniques are: (i) Support Vector Machines (SVMs), (ii) Bayesian Inference, (iii) Dempster-Shafer Theory, (iv) Dynamic Bayesian Networks, (v) Neural Networks (NN) and (vi) Maximum Entropy Model.

Support Vector Machines (SVMs) acquired great popularity for data classification, especially, in the domain of multimedia, where SVM has been used for applications such as face detection, object classification, modality fusion, etc. SVM is a supervised learning method, which assuming a set of input data vectors, provides an optimal binary classification, partitioning the input data into the two training classes. Typically, SVMs are used for multimodal fusion, assuming the set of inputs represents the scores given by individual classifiers. Multimodal fusion and classification using SVMs partitions the input data, applying different kernel functions which allow non-linear classification. Many existing literature approaches use the SVM-based fusion scheme. Nirmala et al. [35] proposed a multimodal image fusion technique using Shift Invariant Discrete Wavelet Transform (SIDWT) for surveillance applications. This approach addressed the fusion of visual and infrared images, extracting their SIDWT and using SVM to fuse the transforms at feature-level. The proposed multimodal image fusion technique combined two information sources, but enabled its extension providing a scalable approach. To compute the SIDWT, images were divided into non-overlapping blocks of fixed size and three features including energy, entropy and standard deviation, were computed for each block. The SVM was trained based on the extracted features for each block and determined whether the wavelet coefficient block from the visual or infrared image was to be used. Finally, the fused image was obtained by performing inverse SIDWT on the selected coefficients. Arsic et al. [3] targeted the automatic detection of certain passenger's behaviour in an airplane situation, e.g., aggressive, nervous, tired, etc, using an SVM during the classification stage. A set of low-level features based on difference imaging were extracted from different parts of the image such as skin colour regions, face or the entire image.

The proposed low-level features were based on the global motion, representing its movement or mean deviation. Finally, a vector containing all features was created and classified using a SVM based on the polynomial kernel.

Bayesian inference combines multimodal information by applying rules of probability theory [27]. Multimodal information sources provide either features or decisions from individual classifiers which are combined to derive the inference of the joint probability of an observation or decision [39]. Bayesian networks allow the use of prior knowledge about the likelihood of the hypothesis to be utilised in the inference process. New observations or decisions can be used to update the a-priori probability in order to compute the posterior probability of the hypothesis. Finally, the Bayesian inference fusion method allows for uncertainty modelling. The Bayesian inference method has been used in the literature to combine multimodal information due to its possibility to adapt as the information evolves as well as its capability to apply subjective or estimated probabilities when empirical data is absent. Due to these advantages, Bayesian inference has been used for different tasks, such as speech recognition or video analysis. In surveillance, Bayesian inference has been applied for combining classification results for various applications [7, 8]. Atrey et al. [6] fused multimodal information using the Bayesian inference fusion approach for event detection in surveillance scenarios, such as standing and talking, running and shouting, walking or standing and door knocking. Meuter et al. addressed vision-based traffic sign recognition in a hybrid classification method based on a decision tree and a Bayesian fusion algorithm [31]. The fusion module combined the classification results of the different classifiers over time and fused similar signs on both sides of the road, taking advantage of the redundancy existent in German roads, where identical signs are mounted on both sides of the road.

Dempster-Shafer Theory allows the inclusion of belief and plausibility values to represent evidences and their corresponding uncertainty in the fusion process, rather than representing the evidence using only uncertainty values [4]. According to the Dempster-Shafer theory, an hypothesis is characterised by belief and plausibility. Whilst the degree of belief implies a lower bound of the confidence, the plausibility represents the upper bound, delimiting the confidence interval or the possibility of the hypothesis to be true [5]. After the assignment of a probability to every hypothesis, the decision regarding the hypothesis is measured by a confidence interval. Multimodal fusion using the Dempster-Shafer theory applies evidence combination rules to fuse multimodal information. Multimodal fusion techniques have acquired great relevance in recent years, the inclusion of higher levels of freedom within the fusion process has been used in different applications. For instance, vehicle classification based on Dempster-Shafer theory was addressed by Klausner et al. [25]. The proposed approach fused single-source classifier's results into a matrix of uncertainty intervals. The authors applied the SVM distance mass function and the Dempster-Shafer belief function to classify objects into three categories, including large trucks, small trucks and cars according to a set of visual and acoustic features. Moreover, Dempster-Shafer theory of evidence was applied on other surveillance applications such as gender profiling. In [28], authors proposed a multimodal fusion technique based on the Dempster-Shafer theory to combine the partial decisions provided by different gender profiling techniques to overcome existing limitations such as the face occlusion or body shape alteration. The provided experiments exhibited an improvement versus single profiling or classic fusion results.

Dynamic Bayesian Networks (DBNs) Multimodal fusion considering the temporal axis requires specific models to describe the evolution of the observed data. For the analysis and fusion of this type of information, Bayesian inference fusion methods can be extended to DBN, also called probabilistic generative model or a graphical model [5]. DBNs have been applied in a diverse range of multimedia applications where the time-series data affected the analysis due to its two main advantages, (i) its ability to model multinode dependency and (ii) to integrate the temporal dependency of the multimodal data. Despite, the variety of DBN systems proposed, the most popular and simplest form of a DBN is the Hidden Markov Model (HMM). HMMs have been used for diverse applications from recognising tennis strokes to gait-based human identification. Additionally, their ability to exploit the spatio-temporal patterns has driven human activity recognition research. A comprehensive review of modelling, recognition and analysis of human activities and interactions was presented by Turaga et al. [44]. Amongst the existing human activity recognition techniques based on DBNs, techniques could be categorised according to the number of agents involved in the activity and/or the amount of information sources. Oliver et al. [36] proposed a system for detection of two person interactions using coupled hidden Markov models (CHMMs). The CHMMs was a variant of HMMs which integrated two or more sources of information to model and recognise human behaviour. Liu and Chua [26] proposed a technique to classify three agent activities, including groups approaching, walking together or meeting and turning back, applying Observation Decomposed Hidden Markov Model (ODHMM). Whilst, Due et al. [15] proposed to decompose an interaction into multiple interacting stochastic processes and proposed a coupled hierarchical durational-state dynamic Bayesian Network. Suk et al. [43] analysed human interactions based on their moving trajectories. Each human interaction was decomposed into elementary components or subinteractions, which were modelled individually using HMMs, and finally assembled using a directed graph. Despite DBNs great development in human activity recognition, multimodal fusion using DBNs was also applied in other surveillance applications such as vision-based traffic monitoring [10, 21], vehicle detection [11], scene description [24, 40] or action recognition based on contextual information [33].

Neural Networks (NN) provide a nonlinear mapping between the input information sources and the output decisions. The NN method consists of a network including input, hidden and output nodes. The input nodes accept information from the different sources while the output nodes provide the results of combining the input information or decisions. The mapping between the input and output nodes, using the hidden nodes, defines the network architecture and therefore its behaviour. The architecture and the weights defining its topology can be adjusted during the training phase to obtain the optimal fusion results [9]. In recent years, multiple applications have used NNs as a multimodal fusion technique. In general scenarios, applications such as speaker tracking [50] or structural damage detection [22] applied NNs to combine different information sources. In surveillance scenarios, NNs have been widely used in traffic control [37]. For example, in [30], traffic magnetic sensors captured the information to detect traffic incidents using NNs. Traffic flow prediction was tackled using a radial basis function neural network [48] or genetic-based NNs [46]. Furthermore, NNs and SVMs were compared for the prediction of traffic speed in [45]. Authors determined that SVM was a viable alternative to NNs for short-term prediction due to the high dependence of NNs performance to the training stage. Despite NNs are suitable for high-dimensional problem spaces and generating high-order nonlinear mapping, NNs present several challenges, including (i) slow training and (ii) complexity to select an appropriate network architecture according to the application under analysis. These challenges limited NNs impact on the multimedia analysis compared to other fusion methods [5].

Maximum Entropy Model presents a statistical classifier which provides a probability of an observation belonging to a particular class based on the input information. The maximum entropy model is used in multimodal fusion, classifying fused multimedia observations, coming from different acquisition sources, into a set of pre-defined concepts. The maximum entropy model-based fusion method learns possible correlations between the extracted features and the selected concepts to build statistical models calculating the probability of the observations belonging to a certain class. The maximum entropy model has been applied for semantic multimedia indexing and annotation. In [29], text and image features were combined to index and retrieve images using the Maximum Entropy Model. The proposed approach was evaluated against the Naives Bayes classifier using the Reuters-21578, Corel Images and TRECVID 2005 datasets. In [2], a multimedia-content automatic annotation approach based on maximum entropy models was presented. Statistical models were calculated extracting colour, texture and shape features to represent each classifying concept. For each concept to be predicted, the set of relevant models were extracted to estimate the probability of the observation to be a particular concept.

7.3 Multimodal Surveillance System

Surveillance systems have been positioned in a fast-track race towards the creation of smart, dynamic, distributed, scalable, multimodal systems. Recent social and technological factors encourage this evolution, including:

- the ever-increasing amount of information collected everyday from the police (from surveillance cameras, police records, speed sensors, etc)
- the in-growing deployment of new privately owned sensors
- the environments variability
- the fast evolution of technological equipment.

The consideration of these factors increases the complexity of surveillance systems tasks, implying an extension of surveillance systems' capabilities to:

- integrate different kinds of sensors

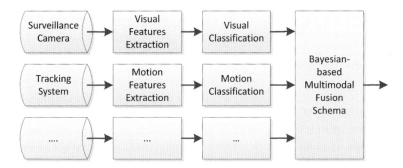

Fig. 7.2 Multimodal surveillance system framework

- process independently each sensor's input
- provide stability handling absence of information or ambiguous inputs
- dynamically expand and adapt its structure and; the complementary information by exploiting the synergy amongst sensors.

Despite several challenges arose from this ambitious system, a crucial research task is the multimodal fusion of information, formally referred as the integration of multiple media, their associated features or the intermediate decision to perform an analysis task. Considering the requirements established for a surveillance system capable to handle realistic situations, a MSS is proposed (refer to Fig. 7.2). The MSS focuses on monitoring urban areas and it is built as a distributed multimodal network. The system, aimed to be scalable and over several multimodal inputs, is proposed as a distributed system, based on the rationale *divide et impera*. Consequently, MSS intends to:

- Decrease the processing time for real-time applications based on distributed decisions.
- Enable scalability, accepting the integration of a non-pre-defined number of inputs during the systems' lifetime.
- Handle uncertainty and lack of information, relying the decisions in the prior-knowledge and the existing evidences, and hence offering a stable operational mode.

The MSS proposes a distributed sensor network, where multimodal fusion is performed at a decision-level. Hence, the individual partial decisions are computed locally and concurrently, reducing the computational load in the centric framework while providing the fusion schema with enough evidences to address high-level analysis. The proposed system is focused on the multimodal classification of moving objects into a set of predefined semantic classes, handing key knowledge to intelligent surveillance systems for more complex context-dependent tasks such as suspicious event detection.

Despite MSS is a scalable surveillance network, real surveillance networks are still not deployed and less connected. Thus, the access to numerous diverse multimodal sensors is a challenging task. In this chapter, the MSS system is built towards the analysis of a specific use case, object classification in urban environments. Considering that object classification can be performed based on a wide range of multimodal information and its fundamental importance as prior step towards suspicious event detection techniques, situates this use case as a critical challenge. The proposed use case analyses two inputs: surveillance videos and tracking coordinates.[1] MSS is a two-stage framework. First, the acquisition systems perform object classification based on partial knowledge of the situation, providing a local-decision and the classifier-confidence level. Second, a Bayesian-based Multimodal fusion schema proceeds with the combination of the partial decisions to maximise the knowledge by exploiting the synergy between complementary information. MSS results with a final classification decision accompanied by a confidence level based on the individual partial decisions, the confidence on the classifiers and the prior knowledge on the urban environment.

In the following paragraphs each MSS stage is further detailed with special attention to the proposed fusion schema.

7.3.1 MSS Multimodal Inputs

The complexity inherent in urban surveillance requires collaborative work from different sensors to enable certain capabilities such as the detection of suspicious events, despite the external factor affecting the scene (refer to Fig. 7.3). Based on this idea, MSS proposes to divide the information to analyse and obtain individual remote decisions to build a knowledge network provider. Hence, two inputs are feeding the MSS, surveillance videos and tracking coordinates.

7.3.1.1 Surveillance Videos

A moving object classifier is presented based on the analysis of the CCTV videos and the extraction and modelling of visual patterns [17]. The surveillance centric object classifier consists of three stages:

- Motion analysis component targets the extraction of moving objects in the surveillance videos to optimise the next stages performance by selecting relevant information, removing the irrelevant and so reducing the computational burden. This stage targets background subtraction based on Gaussian Mixture Models, moving object segmentation using connected components and object tracking applying Kalman filters to predict the tracks.
- Feature extraction component addresses the creation of visual patterns for each segmented moving object in the video. Different low-level features were analysed,

---

[1] Despite the limited number of inputs, MSS is built in a scalable fashion, prepared to aggregate new inputs through the system's lifetime.

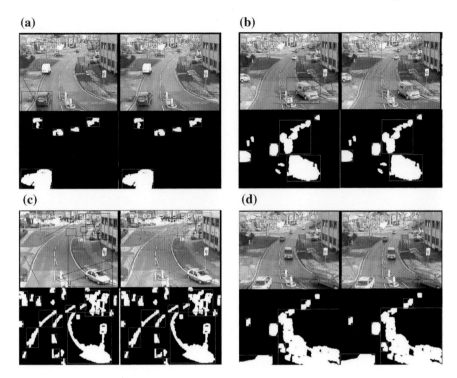

Fig. 7.3 Video analysis techniques' results are affected by several problematic situations as low quality image (**a**), inaccurate background subtraction (**b**), camera movement (**c**) and objects merged due to noise and shadows (**d**). The four images composing this figure (**a–d**) show the original images (in colour and *black & white*, *top left* and *top right* images, respectively) as well as the results obtained after processing the surveillance videos by video analysis techniques, i.e., background subtraction, object detection and object tracking (*bottom images*). The result images reflect the effect of external factors on the obtained results

including global and local descriptors. However, after an exhaustive analysis only four of them were selected due to their high distinctiveness, compact representation and significance for human perception. These features are the following MPEG-7 descriptors: Colour Layout Descriptor (CLD), Edge Histogram Descriptor (EHD), Dominant Colour Descriptor (DCD) and Colour Structure Descriptor (CSD). All these features were chosen for their representativeness for the visual patterns useful in surveillance applications (for further information on the visual feature performance comparison and analysis for surveillance applications refer to [18]).

- Multifeature fusion algorithm. Based on the assumption that single visual feature descriptors are not capable of interpreting human understanding, a combination of visual features could represent more complex patterns. However, each feature has different nature, metrics and nonlinear behaviour. Hence, to combine single features, these requirements must be considered. In this visual-based object classifier, the feature combination relies on the Multi-Objective Optimisation technique (MOO) [49]. The main idea is to perform a weighted average feature combination where weights are optimised and features are analysed and compared in their natural feature space. These weights are optimised through the use of pareto optimal solutions, which minimise the distance between the object visual features and the centroid of the feature's positive training samples and maximising it with the negative training examples. This rationale is resumed in:

$$\min \left\{ \frac{\sum_{k=1}^{K} D_+^{(k)}(V^{(k)}, \bar{V}, A_s)}{\sum_{k=1}^{K} D_-^{(k)}(V^{(k)}, \bar{V}, A_s)} \right\}, s = 1, 2, \ldots, S \qquad (7.1)$$

where $D_-^{(k)}$ and $D_+^{(k)}$ are the distances over positive and negative training samples respectively, $V^{(k)}$ represents the visual features vector for the moving object (k), $\bar{V}$ is the features centroid vector formed by the selected MPEG-7 features and represented as $\bar{V} = (\bar{v}_{CLD}, \bar{v}_{SCD}, \bar{v}_{DCD}, \bar{v}_{EHD})$, while, A_s is the sth in the set of *Pareto-optimal solutions*, and S is the number of available *Pareto-optimal solutions* [17].

Finally, this classifier exploits the inherent visual appearance of the moving objects to create a unique and optimal representation vector.

7.3.1.2 Tracking Coordinates

In urban environments, surveillance cameras are not the only sensors deployed, though they are the most common. Other sensors recording global positioning or spatio-temporal evolution of moving objects are also inserted in surveillance distributed networks. The second input to the MSS system provides the tracking coordinates of moving objects located in the same area monitored by the CCTV camera (first MSS input).

Considering the conclusions extracted by certain psychological studies, determining the importance of motion as a fundamental cue for humans to classify objects [23], in this section, a behaviour-based classifier is presented. The main objective is to extract and model behaviour patterns of the moving objects in order to use their spatio-temporal evolution to perform semantic classification. The process consists of two stages:

- Behaviour patterns extraction. A set of features exploiting the intra-object variance while minimising the inter-object variance was selected to represent the behaviour of moving objects, including, *shape pattern*, *velocity* and *trajectory* [16]. Geometric computation algorithms were developed to model object's behaviour and extracted to form the *Behaviour Pattern (BP)*, where
  $BP = \{Shape, Size, Velocity, Trajectory\}$.
- Behaviour fuzzy classification proposes a hierarchical fuzzy classifier based on a general framework consisting of two levels of classification performed on cas-

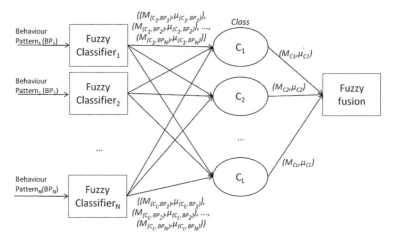

Fig. 7.4 Behavioural fuzzy classifier framework

cade to provide robustness against behaviour pattern outliers (shown in Fig. 7.4[2]). The first level depicts the classification of each moving object according to every individual behaviour feature. This level is built with a set of nested rule-based fuzzy classifiers. The membership functions applied for each behaviour feature are extrapolated from the marginal training sample created from the manually annotated dataset. As a result, each individual fuzzy classifier provides a classification label and a membership degree. While the second level of the hierarchical fuzzy classifier performs the combination of the individual classification results obtained in the first level, through a set of high-level fuzzy classification rules.

Finally, this classifier exploits the spatio-temporal evolution of the moving objects to create a unique and optimal representation vector depicting its evolutionary behaviour.

7.3.2 Bayesian-Based Multimodal Fusion Technique

The diverse nature of the sensors deployed in a controlled area requires a versatile and adaptive fusion technique, capable to overcome their individual challenges, as well as, address the absence of information and the presence of uncertainty by the means of inferring information from previously acquired knowledge. Hence, in this chapter, we propose a Bayesian-based Multimodal Fusion technique to provide a

---

[2] In the figure, M_x is the label which indicates if certain behaviour feature, BF_j, fulfils the condition attached to the semantic class, C_x, and $CF = \mu_{jx}$ represents the membership degree of the behaviour feature under analysis to belong to the semantic class C_x, μ represents the membership degree and N is the amount of behaviour features considered in the analysis.

7 Multimodal Fusion in Surveillance Applications

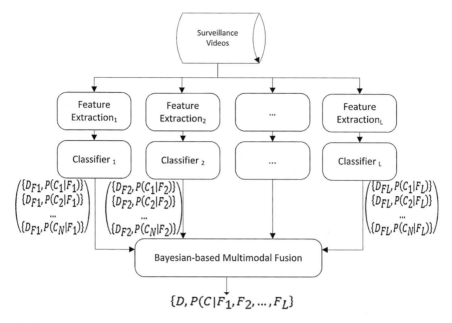

Fig. 7.5 Bayesian-based multimodal fusion technique for object classification in surveillance networks

scalable algorithm capable to combine the two aforementioned inputs (refer to the scalable framework presented in Fig. 7.5).

The rationale to use Bayesian Networks relies on its advantages for integration and robustness based on the application of probability theory. There are three main advantages characteristic from Bayesian Networks [19]. First, the Bayesian inference method allows the combination of multimodal information due to its possibility of adaptation as the information evolves as well as its capability to apply subjective or estimated probabilities when empirical data is absent [5]. Second, the hierarchical structure provides flexibility and scalability, facilitating not only the inclusion of additional information, but also enabling the degradation of the a-posteriori probability in case of the absence of a certain cue/s. Finally, Bayesian networks allow domain knowledge to be embedded in the structure and parameters of the networks, allowing the adjustment of the fusion technique to the domain and scenario's requirements. Based on these advantages, the proposed Bayesian-based Multimodal Fusion technique provides the following benefits to the MSS system:

- Combines the output from several diverse-nature sensors exploiting their complementary.
- Addresses absence of information basing the classification on the input received and the prior knowledge and decreasing the certainty of the classification if evidences are missing.

- Each sensor performs local-based classification, alleviating the computational requirements of the central MSS system.
- Provides scalability based on the Bayesian Networks adaptability advantage.

Each of the aforementioned individual classifiers provides a partial decision and a conditional probability matrix describing the probability of a detected moving object to belong to each of the semantic concepts, $C_i, i = 1, \ldots, N$, considered within the classification scenario, and the matrices are defined as:

$$\begin{pmatrix} P(C_1|F_1) & P(C_1|F_2) \\ P(C_2|F_1) & P(C_2|F_2) \\ \ldots & \ldots \\ P(C_N|F_1) & P(C_N|F_2) \end{pmatrix} \tag{7.2}$$

where F_j represents each of the individual classifiers whose decisions are fused applying Bayes' probabilistic rules. Each partial decision could perform automatic object classification. However, the integration of several features, derived from different and uncorrelated media, addresses higher robustness, stability, flexibility and adaptation.

The proposed Bayesian-based multimodal fusion approach performs probabilistic fusion at a decision-level, standardising the output of the remote local classifiers and thus enabling distributed classification for surveillance networks. Considering the decision-level fusion as a classification problem, the Bayesian inference scheme can be formulated using the maximum a-posteriori criterion (MAP):

$$D = \underset{i}{\operatorname{argmax}}\{P(C_i|F_1, F_2, \ldots, F_L)\} = \underset{i}{\operatorname{argmax}}\left\{\prod_{j=1}^{L} P(F_j|C_i)P(C_i)\right\}$$

$$= \underset{i}{\operatorname{argmax}} \begin{pmatrix} \prod_{j=1}^{L} P(F_j|C_1)P(C_1) \\ \prod_{j=1}^{L} P(F_j|C_2)P(C_2) \\ \vdots \\ \prod_{j=1}^{L} P(F_j|C_N)P(C_N) \end{pmatrix} \tag{7.3}$$

where $P(C_i|F_1, F_2, \ldots, F_L)$ defines the probability of a concept C_i to be the final decision undertaken by the classifier, D, considering all the individual partial decisions provided by individual classifiers; F_j are the individual classifiers that provide partial decisions to the Bayesian inference scheme; $P(C_i)$ represents the a-priori probability of the i concept; L defines the amount of partial decisions incorporated in the multimodal fusion, which in MSS this implies $L = 2$, and N represents the number of concepts involved in the classification problem (described in Sect. 7.3.1). Finally, the conditional probability matrices connecting distributed partial decisions to the surveillance network can be set manually or learned from training data, and so permits adaptation to scenario, application and the topology of the surveillance network.

The proposed Bayesian-based Multimodal Fusion technique, core of the Multimodal Surveillance System, presents a semantic classification technique based on the fusion of diverse-nature cues. The semantic object classification provides several advantages to future applications including (i) to enhance forensic applications enabling the hyper-connexion between the automatic classification, the queries and concepts meaningful for human operators and (ii) for the detection of surveillance events in urban environments, the use of human understanding in the decision-making process enables the capability to establish human related rules to infer object-oriented events.

7.4 Performance Evaluation of the Multimodal Surveillance System

During this section, a qualitative and quantitative performance evaluation of the Bayesian-based Multimodal Fusion technique is presented along with a comparison of the proposed individual object classifiers performance. Moreover, the surveillance dataset and ground truth are explained for further understanding of the experiments.

7.4.1 Dataset and Ground Truth

In order to evaluate the performance of the proposed Multimodal Surveillance System and the benefits from combining complementary information from diverse-nature sensors, the Bayesian-based Multimodal Fusion technique was applied to a set of video sequences and tracking data.

AVSS 2007 dataset[3] was used to evaluate the proposed system, providing outdoor videos summing a total of 13,400 images, with variable lighting conditions as well as different levels of difficulty. The surveillance footage includes several challenges such as noise, low quality image, camera movement or blurring increasing the difficulty of its analysis. From these videos, the tracking coordinates from each of the moving objects recorded on the scenes are stored in independent files.

In order to perform careful evaluation, a ground truth was manually annotated, containing the two most common semantic categories within the videos, namely *Person* and *Vehicle*. A total of 1,567 objects were included, 6% were person while 50% were vehicle. Due to the imposed guidelines for the manual annotation and the challenges introduced by the motion analysis component (in the visual-based object classifier), objects presenting certain constraints such as small blob size, partial occlusion of the object over 50% or multiple objects coexisting in a blob, were annotated as *unknown*. The proposed system, MSS, probabilistically categorises

---

[3] http://www.eecs.qmul.ac.uk/~andrea/avss2007_d.html

Table 7.1 Performance comparison of the four selected MPEG-7 features, the linearly combined multi-feature descriptor and the proposed optimal multi-feature descriptor

Semantic concepts	F-measure (%)					
	CLD	EHD	CSD	DCD	SVM	MOO
Person	4.47	19.14	7.76	63.82	7.92	25.35
Vehicle	5.92	68.48	65.38	7.27	45.69	64.43

each moving object into one of those semantic categories, based on independent partial decision provided by the two object classifiers presented in Sect. 7.3.1.

7.4.2 Experimental Results

In order to provide a comprehensive evaluation of the Multimodal Surveillance System and the benefits from exploiting the synergies between multimodal inputs, an individual evaluation of each input detailed in Sect. 7.3.1 is presented, followed by the analysis of the Bayesian-based Multimodal Fusion technique. This Section ends with a comparison amongst the individual decision-making processes.

7.4.2.1 Object Classification Based on the Analysis of Surveillance Videos

In this Section, the proposed Visual-based Object Classifier is evaluated against state-of-the-art machine learning techniques, for individual descriptors as well as for linearly combined multi-feature descriptors. The selected visual features, CLD, EHD, DCD and CSD, were computed to represent and classify all the detected moving objects within the surveillance video datasets according to their visual appearance. The proposed object classifier applies MOO technique to optimise the weighted linear combination for the visual feature descriptors while considering that each feature has a different feature space. The benefits addressed by the MOO technique are further contrasted against SVM in individual feature spaces as well as in multi-feature descriptor spaces built by concatenating the appearance descriptors. Two are the objectives of this comparison: (i) to demonstrate the benefits of multifeature descriptors versus individual descriptors and (ii) to establish the need to preserve each individual feature space while combining visual features. The obtained results are shown in Table 7.1, where six results are presented:

- Each visual feature individual performance, namely CLD, EHD, CSD and DCD.
- A linear concatenation of the four selected visual features to build a multi-feature descriptor, namely SVM.
- The proposed multi-feature visual pattern built based on the MOO technique and preserving the nonlinearity of each individual visual feature, namely MOO.

Table 7.2 Performance of the proposed object classifier based on tracking information analysis

Semantic class	True positive (%)	True negative (%)	False positive (%)	False negative (%)	F-measure
Vehicle	79.40	51.06	48.94	20.60	61.8
Person	51.06	79.40	20.60	48.94	59.5

The results provided by linearly combining several features (SVM) reveal a considerable F-measure value for the concept Vehicle, however, drops its performance for the concept Person. The proposed visual-based object classifier (MOO) show a reasonable improvement for both semantic concepts, Vehicle and Person, outperforming SVM by 18 %, and exceeding single visual features performance, demonstrating the necessity to consider each feature individually respecting its nature, behaviour and specific metrics.[4]

7.4.2.2 Object Classification Based on the Analysis of Tracking Information

The proposed object classifier based on the analysis of the spatio-temporal evolution of the moving objects in a scene provides not only a membership label for each detected moving object, but also a membership degree exhibiting the reliability on the membership label (refer to Sect. 7.3.1.2). However, in order to study the classification results, we strictly consider the membership label to calculate the percentage of false positives, false negatives, true positives and true negatives for each semantic class (refer to Table 7.2).

In Table 7.2, the flexibility provided by fuzzy logic is omitted to study the performance obtained by sharp binary classifiers in a behaviour-based object classifier. The results reveal a high true positive rate in both semantic concepts but affected by a significant false negative rate. The results obtained for the semantic concept Person can be related to the sparseness of this object category within the ground truth, generating a less accurate model. On the other hand, the results for the semantic concept Vehicle are limited for two reasons. First, the analysed scene records a road with urban speed limitation so vehicles do not exceed person's speed, limiting the discriminative effect of the velocity pattern. Second, the appearance of vehicles with different silhouettes and shape ratios, some of them really similar to a person's shape ratio.

7.4.2.3 Evaluation of the Bayesian-Based Multimodal Fusion Technique

In this chapter, a Bayesian-based Multimodal Fusion technique is proposed to classify moving objects according to multiple diverse-nature information, exploiting the

---

[4] Person results can be related to the sparseness of the concept within the ground truth (refer to Sect. 7.4.1).

Table 7.3 Performance evaluation of the bayesian-based multimodal fusion technique

Concepts	True positive (%)	True negative (%)	False positive (%)	False negative (%)	F-measure (%)
Vehicle	97	66	34	3	83.9
Person	66	97	3	34	78.1

synergies amongst complementary information. The proposed fusion technique is performed at a decision-level, enabling the creation of distributed surveillance networks, where each node performs individual classification, providing the central system (MSS) with a partial classification and the confidence on the classification results. The combination of multimodal information in a decision-level fusion technique based on Bayesian Networks enables the working continuity of the MSS system despite the absence of information and the presence of uncertainty.

To evaluate the performance of the proposed Bayesian-based Multimodal Fusion technique, a conditional probability matrix is calculated by each of the individual classifiers, acting as inputs to the Bayesian Network. The Bayesian-based Multimodal Fusion technique combines the locally computed partial decisions to achieve a unique classification, preserving their individual feature spaces and metrics. The obtained results are shown in Table 7.3.

The obtained results reveal a high rate of true positive and low false negative rates for the semantic concept Vehicle, 97 and 3 %, respectively. The semantic concept Person presents lower true positive and false negative rates, 66 and 34 %, respectively. The achieved false positive detection rates are 34 and 3 % for Vehicle and Person, respectively. The results, both the false positive rates for vehicles and the true positive rate for person, are directly affected by the sparseness of the concept person within the ground truth.

7.4.2.4 Comparative Evaluation

The diversity and variability of sensors deployed in surveillance networks entailed a fusion challenge. In this chapter, a Bayesian-based Multimodal Fusion technique was proposed to provide a robust, versatile, scalable solution capable of handling absence of information and presence of uncertainty. The objective of the proposed multimodal fusion technique was to allow the integration of various different nature features independently of which media were they derived from, to benefit from (i) the representability provided by each feature, (ii) their uncorrelation in order to cover a bigger spectrum, and (iii) the robustness acquired by the system due to the consideration of multiple partial decisions rather than relying on a single decision. In order to demonstrate such an improvement on the performance, the Bayesian-based Multimodal Fusion technique is compared with the individual classifiers, presenting the partial decisions as inputs to the Bayesian inference scheme. Table 7.4 presents the comparative results, revealing that the combination of multimodal inputs based on a Bayesian inference scheme outperforms both individual object classifiers. While

Table 7.4 Performance comparison between the proposed bayesian-based multimodal fusion technique and the two intermediate object classifiers based on visual and spatio-temporal features, which are further explained in Sects. 7.3.1.1 and 7.3.1.2, respectively

Concepts		True positive (%)	True negative (%)	False positive (%)	False negative (%)	F-measure (%)
Vehicle	Visual features	77	64	36	23	64.43
	Spatio-temporal features	79	57	43	21	61.8
	Bayesian	97	66	34	3	83.9
Person	Visual features	64	77	23	36	25.35
	Spatio-temporal features	57	79	21	43	59.5
	Bayesian	66	97	3	34	78.1

visual and spatio-temporal-based classifiers achieve a true positive rate of 77 and 79 %, respectively, the Bayesian-based Multimodal Fusion technique reveals a 97 % positive rate for the semantic concept Vehicle. A lower increase of the performance is also registered for semantic concept Person, increasing the true positive rate by 1 and 9 % for the visual and spatio-temporal features classifiers. A reason for the reduced improvement and generally the lower true positive rates obtained for the semantic concept Person is its sparseness in the ground truth and the dataset (for further information refer to Sect. 7.4.1).

Generally, classifiers' performance is represented by their F-measure. Considering this measurement, the proposed Bayesian-based Multimodal Fusion technique reveals the benefits of exploiting the synergies existing between complementary information, exceeding with its results both individual classifiers by 18.57 and 22.1 % respectively for the semantic concept Vehicle and by 52.75 and 18.6 % respectively for the semantic concept Person. A refined analysis reveals that the proposed Bayesian-based Multimodal Fusion technique enhances the object classification procedure, increasing positive detection and reducing false alarms.

7.5 Conclusions and Future Work

In this book chapter, the foundations for a MSS envisaged as a distributed network of sensors is presented, together with a probabilistic fusion schema proposed as the core of the centric system. For the evaluation of the MSS, its implementation and experiments have been implemented analysing a case of study, object classification in urban environments. Two inputs fed the MSS, surveillance videos and tracking information, and were computed locally, providing the partial classification results and the confidence level of the classifier to the Bayesian-based Multimodal Fusion component. The Bayesian inference schema, based on the premise that the higher amount of complementary information would provide higher robustness and accuracy in the decision-making process, exploited the synergies of the two independent inputs, obtaining higher rate of true positive detections and lower false alarms than the partial decisions, and generally, higher F-measure results. The proposed fusion technique provided the foundation to build a distributed, scalable, dynamic multimodal surveillance network, addressing also the partial or total absence of information, by degrading the classification results accordingly. The proposed Bayesian-based Multimodal Fusion technique outperformed both individual classifiers, demonstrating the benefits of combining complementary features to improve the classification results and to enhance the robustness of the classification framework.

The use of the proposed technique as prior step for event detection in urban environments will be addressed in the future. Moreover, in this book chapter, a challenging research line was presented. In the future, we will target the consolidation of a multimodal surveillance network, fed by numerous inputs and favoured with a inference schema capable to extract inherent information for urban scene monitoring.

Acknowledgments This research was partially supported by the European Commission under contract FP7-261743 VideoSense.

References

1. Aghajan H, Cavallaro A (2009) Multi-camera networks: principles and applications. Academic Press, London
2. Argillander J, Iyengar G, Nock H (2005) Semantic annotation of multimedia using maximum entropy models. In: IEEE international conference on acoustics, speech, and signal processing, vol 2. pp 153–156
3. Arsic D, Schuller B, Rigoll G (2007) Suspicious behavior detection in public transport by fusion of low-level video descriptors. In: IEEE international conference on multimedia and expo, pp 2018–2021
4. Atrey P, Kankanhalli M, El Saddik A (2006) Confidence building among correlated streams in multimedia surveillance systems. Adv Multimedia Model 4352:155–164
5. Atrey PK, Hossain MA, El Saddik A, Kankanhalli MS (2010) Multimodal fusion for multimedia analysis: a survey. Multimedia Syst 16(6):345–379
6. Atrey PK, Kankanhalli MS, Jain R (2006) Information assimilation framework for event detection in multimedia surveillance systems. Multimedia syst 12(3):239–253
7. Bahlmann C, Pellkofer M, Giebel J, Baratoff G (2008) Multi-modal speed limit assistants: Combining camera and gps maps. In: IEEE intelligent vehicles symposium, pp 132–137
8. Bahlmann C, Zhu Y, Ramesh V, Pellkofer M, Koehler T (2005) A system for traffic sign detection, tracking, and recognition using color, shape, and motion information. In: IEEE intelligent vehicles symposium, pp 255–260
9. Brooks RR, Iyengar SS (1998) Multi-sensor fusion: fundamentals and applications with software. Prentice-Hall Inc, Upper Saddle River
10. Buxton H, Gong S (1995) Visual surveillance in a dynamic and uncertain world. Artif Intell 78(1–2):431–459
11. Cheng HY, Weng CC, Chen YY (2012) Vehicle detection in aerial surveillance using dynamic bayesian networks. IEEE Trans Image Process 21(4):2152–2159

12. Csurka G, Clinchant S. An empirical study of fusion operators for multimodal image retrieval. In: 10th international workshop on content-based multimedia indexing, IEEE, pp 1–6
13. Dore A, Pinasco M, Regazzoni C (2009) Multi-modal data fusion techniques and applications. In: Multi-camera networks, pp 213–237
14. Drajic D, Cvejic N (2007) Adaptive fusion of multimodal surveillance image sequences in visual sensor networks. IEEE Trans Consum Electron 53(4):1456–1462
15. Du Y, Chen F, Xu W (2007) Human interaction representation and recognition through motion decomposition. IEEE Sig Process Lett 14(12):952–955
16. Fernandez Arguedas V, Izquierdo E (2011) Object classification based on behaviour patterns. In: International conference on imaging for crime detection and prevention
17. Fernandez Arguedas V, Zhang Q, Chandramouli K, Izquierdo E (2011) Multi-feature fusion for surveillance video indexing. In: International workshop on image analysis for multimedia interactive services, IEEE
18. Fernandez Arguedas V, Zhang Q, Chandramouli K, Izquierdo E (2012) Semantic hyper/multi media adaptation, chapter Vision based semantic analysis of surveillance videos. Springer, Berlin, pp 83–126
19. Fernandez Arguedas V, Zhang Q, Izquierdo E (2012) Bayesian multimodal fusion in forensic applications. In: Computer vision-ECCV 2012, workshops and demonstrations. Springer, pp 466–475
20. Gupta H, Yu L, Hakeem A, Eun Choe T, Haering N, Locasto M (2011) Multimodal complex event detection framework for wide area surveillance. In: IEEE computer society conference on computer vision and pattern recognition workshops, pp 47–54
21. Huang T, Koller D, Malik J, Ogasawara G, Rao B, Russell S, Weber J (1995) Automatic symbolic traffic scene analysis using belief networks. In: Proceedings of the national conference on artificial intelligence. Wiley, pp 966–966
22. Jiang SF, Zhang CM, Zhang S (2011) Two-stage structural damage detection using fuzzy neural networks and data fusion techniques. Expert Syst Appl 38(1):511–519
23. Johansson G (1973) Visual perception of biological motion and a model for its analysis. Attention Percept Psychophys 14(2):201–211
24. Junejo IN (2010) Using dynamic bayesian network for scene modeling and anomaly detection. Sig Image Video Process 4(1):1–10
25. Klausner A, Tengg A, Rinner B (2007) Vehicle classification on multi-sensor smart cameras using feature-and decision-fusion. In: ACM/IEEE international conference on distributed smart cameras, pp 67–74
26. Liu X, Chua CS (2006) Multi-agent activity recognition using observation decomposedhidden markov models. Image Vis Comput 24(2):166–175
27. Luo RC, Yih CC, Su KL (2002) Multisensor fusion and integration: approaches, applications, and future research directions. IEEE Sens J 2(2):107–119
28. Ma J, Liu W, Miller P (2012) An evidential improvement for gender profiling. In: Denoeux T, Masson M-H (eds) Belief functions: theory and applications, volume 164 of advances in intelligent and soft computing. Springer, Berlin/Heidelberg, pp 29–36
29. Magalhães J, Rüger S (2007) Information-theoretic semantic multimedia indexing. In: ACM international conference on image and video retrieval, pp 619–626
30. Messai N, Thomas P, Lefebvre D, Moudni AE (2005) Neural networks for local monitoring of traffic magnetic sensors. Control Eng Pract 13(1):67–80
31. Meuter M, Nunn C, Görmer SM, Müller-Schneiders S, Kummert A (2011) A decision fusion and reasoning module for a traffic sign recognition system. IEEE Trans Intell Transp Sys 99:1–9
32. Mironica I, Ionescu B, Knees P, Lambert P (2013) An in-depth evaluation of multimodal video genre categorization. In: 11th International workshop on content-based multimedia indexing, pp 11–16
33. Moore DJ, Essa IA, Hayes MH III (1999) Exploiting human actions and object context for recognition tasks. In: IEEE international conference on computer vision, vol 1. pp 80–86
34. Nayak J, Gonzalez-Argueta L, Song B, Roy-Chowdhury A, Tuncel E (2008) Multi-target tracking through opportunistic camera control in a resource constrained multimodal sensor

network. In: Second ACM/IEEE international conference on distributed smart cameras, ICDSC 2008, pp 1–10
35. Nirmala DE, Paul BS, Vaidehi V (2011) A novel multimodal image fusion method using shift invariant discrete wavelet transform and support vector machines. In: International conference on recent trends in information technology, pp 932–937
36. Oliver NM, Rosario B, Pentland AP (2000) A bayesian computer vision system for modeling human interactions. IEEE Trans Pattern Anal Mach Intell 22(8):831–843
37. Ozkurt C, Camci F (2010) Automatic traffic density estimation and vehicle classification for traffic surveillance systems using neural networks. Math Comput Appl 14(3):187
38. Prati A, Vezzani R, Benini L, Farella E, Zappi P (2005) An integrated multi-modal sensor network for video surveillance. In: ACM international workshop on video surveillance and sensor networks, pp 95–102
39. Rashidi A, Ghassemian H (2003) Extended dempster-shafer theory for multi-system/sensor decision fusion. In: Joint workshop on challenges in geospatial analysis, integration and visualization II, pp 31–37
40. Remagnino P, Tan T, Baker K (1998) Agent orientated annotation in model based visual surveillance. In: International conference on computer vision, pp 857–862
41. Snidaro L, Visentini I, Foresti G (2011) Data fusion in modern surveillance. In: Innovations in defence support systems-3, pp 1–21
42. Snoek CGM, Worring M, Smeulders AWM (2005) Early versus late fusion in semantic video analysis. In: ACM international conference on multimedia
43. Suk HI, Jain AK, Lee SW (2011) A network of dynamic probabilistic models for human interaction analysis. IEEE Trans Circ Syst Video Technol 21(7):932–945
44. Turaga P, Chellappa R, Subrahmanian VS, Udrea O (2008) Machine recognition of human activities: a survey. IEEE Trans Circ Sys Video Technol 18(11):1473–1488
45. Vanajakshi L, Rilett LR (2004) A comparison of the performance of artificial neural networks and support vector machines for the prediction of traffic speed. In: IEEE intelligent vehicles symposium, pp 194–199
46. Vlahogianni EI, Karlaftis MG, Golias JC (2005) Optimized and meta-optimized neural networks for short-term traffic flow prediction: a genetic approach. Transp Res Part C Emerg Technol 13(3):211–234
47. Wu Z, Cai L, Meng H (2006) Multi-level fusion of audio and visual features for speaker identification. In: International conference on advances in biometrics, pp 493–499
48. Xiao JM, Wang XH (2004) Study on traffic flow prediction using rbf neural network. In: International conference on machine learning and cybernetics, vol 5. pp 2672–2675
49. Zhang Q, Izquierdo E (2007) Combining low-level features for semantic inference in image retrieval. In: EURASIP journal on advances in signal processing, p 12
50. Zou X, Bhanu B (2005) Tracking humans using multi-modal fusion. In: IEEE computer society conference on computer vision and pattern recognition, pp 4–4

Chapter 8
Multimodal Violence Detection in Hollywood Movies: State-of-the-Art and Benchmarking

Claire-Hélène Demarty, Cédric Penet, Bogdan Ionescu, Guillaume Gravier and Mohammad Soleymani

Abstract This chapter introduces a benchmark evaluation targeting the detection of violent scenes in Hollywood movies. The evaluation was implemented in 2011 and 2012 as an affect task in the framework of the international MediaEval benchmark initiative. We report on these 2 years of evaluation, providing a detailed description of the dataset created, describing the state of the art by studying the results achieved by participants and providing a detailed analysis of two of the best performing multimodal systems. We elaborate on the lessons learned after 2 years to provide insights on future work emphasizing multimodal modeling and fusion.

8.1 Introduction

Detecting violent scenes in movies appears as an important feature in various use cases related to video on demand and child protection against offensive content. In the framework of the MediaEval benchmark initiative, we have developed a large dataset for this task and assessed various approaches via comparative evaluations.

C.-H. Demarty (✉) · C. Penet
Technicolor, 975 av. des Champs Blancs, 35576 Cesson Sévigné Cedex, France
e-mail: claire-helene.demarty@technicolor.com

C. Penet
e-mail: penetcedric@gmail.com

B. Ionescu
LAPI, University Politehnica of Bucharest, 061071 Bucharest, Romania
e-mail: bionescu@imag.pub.ro

G. Gravier
IRISA and INRIA Rennes, 35042 Rennes Cedex, France
e-mail: guig@irisa.fr

M. Soleymani
iBUG, Imperial College London, London SW7 2AZ, UK
e-mail: m.soleymani@imperial.ac.uk

MediaEval[1] is a benchmarking initiative dedicated to evaluating new algorithms for multimedia access and retrieval. MediaEval emphasizes the multimodal character of the data (speech, audio, visual content, tags, users, context, etc). As a track of MediaEval, the Affect Task—Violent Scenes Detection—involves automatic detection of violent segments in movies. The challenge derives from a use case at the company Technicolor.[2] Technicolor is a provider of services in multimedia entertainment and solutions, in particular, in the field of helping users select the most appropriate content according to, for example, their profile. In this context, a particular use case arises which involves helping users choose movies that are suitable for children in their family, by previewing the parts of the movies (i.e., scenes or segments) that include the most violent moments [9].

Such a use case raises several substantial difficulties. Among them, the subjectivity that will occur during the selection of those violent moments is certainly the most important one. Indeed the definition of a violent event remains highly subjective and dependent on the viewers, their culture, their gender. Agreeing on a common definition of a violent event is not easy, which explains why each work related to violence in the literature exhibits a different definition. The semantic nature of the events to retrieve also contributes to the difficulty of the task, as it entails a huge semantic gap between features and interpretation. Due to the targeted content (i.e., Hollywood movies) and the nature of the events, multimodality is also an important characteristic of the task, which stresses its ambitious and challenging nature even more.

The choice of the targeted content raises additional challenges which are not addressed in similar evaluation tasks, for example in the TRECVid Surveillance Event Detection or Multimedia Event Detection Evaluation Tracks.[3] Indeed, systems will have to cope with content of very different genres that may contain special editing effects, which may alter the events to detect.

In the literature, violent scene detection in movies has received very little attention so far. Moreover, comparing existing results is impossible because of the different definitions of violence adopted. As a consequence of the differences in the definition of violence, methods suffer from a lack of standard, consistent, and substantial datasets. The Affect task of MediaEval constitutes a first attempt to address all these needs and establish a standard with state-of-the-art performance for future reference.

This paper provides a thorough description of the Violent Scene Detection (VSD) dataset and reviews the state of the art for this task. The main contributions in this regard can be summarized with:

- the proposal of a definition of violence in movies and its validation in the community,
- the design of a comprehensive dataset of 18 Hollywood movies annotated for violence and for concepts related to violence. Insights about annotation challenges are also provided;

---

[1] http://www.multimediaeval.org/

[2] http://www.technicolor.com/

[3] http://www.nist.gov/itl/iad/mig/sed.cfm

- a detailed description of the state of the art in violence detection;
- a comparison of the systems that competed in the 2011 and 2012 benchmarks and the description of two of the best performing systems.

The chapter is organized as follows. Section 8.2 reviews previous research on violence detection in videos. Section 8.3 provides an overview of the violent scene detection task after 2 years of implementation within the MediaEval benchmarking initiative. Section 8.4 reports the results of the benchmark with a short comparative description of the competing systems. Section 8.5 provides an in-depth description of two of the best ranked systems with an explicit focus on the contribution of the multimodal information fusion.

8.2 A Review of the Literature

Automatically detecting violent scenes in movies received very limited attention prior to the establishment of the MediaEval violence detection task [21].

A closely related problem is action recognition focusing on detecting human violence in real-world scenarios. Datta et al. [8] proposed an hierarchical approach for detecting distinct violent events involving two people, e.g., fist fighting, hitting with objects, and kicking. They computed the motion trajectory of image structures, i.e., acceleration measure vector and its jerk. Their method was validated on 15 short sequences including around 40 violent scenes. Another example is the approach in [40] which aims at detecting instances of aggressive human behavior in public environments. The authors used a Dynamic Bayesian Network (DBN) as a fusion mechanism to aggregate aggression scene indicators, e.g., "scream," "passing train," or "articulation energy." Evaluation is carried out using 13 clips featuring various scenarios, such as "aggression towards a vending machine" or "supporters harassing a passenger."

Sports videos were also used for violence detection, usually relying on the bag of visual words (BoVW) representation. For instance, [32] addresses fight detection using BoVW along with space-time interest points and motion scale-invariant feature transform (MoSIFT) features. The authors evaluated their method on 1,000 clips containing different actions from ice hockey videos labeled at the frame level. The highest reported detection accuracy is near 90 %. A similar experiment is the one in [11] that used BoVW with local spatio-temporal features, for sports and surveillance videos. Experiments show that motion patterns tend to provide better performance than spatio-visual descriptors.

One of the early approaches targeting broadcast videos is from Nam et al. [31] where violent events were detected using multiple audio–visual signatures, e.g., description of motion activity, blood and flame detection, and violence/nonviolence classification of the soundtrack and characterization of sound effects. Only qualitative validations were reported. More recently, Gong et al. [17] used shot length, motion activity, loudness, speech, light, and music as features for violence detection. A modified semi-supervised learning model was employed for detection and evaluated on 4 Hollywood movies, achieving a F-measure of 0.85 at best. Similarly, Giannakopoulos et al. [14] used various audio-visual features for violence detection in movies, e.g., spectrogram, chroma, energy entropy, Mel-Frequency Cepstral Coefficients (MFCC), average motion, motion orientation variance, measure of the motion of people or faces in the scene. Modalities were combined by a meta-classification architecture that classified mid-term video segments as "violent" or "non-violent." Experimental validation was performed on 50 video segments ripped from 10 different movies (totaling 150 min) with F-measures up to 0.58. Lin and Wang [27] proposed a violent shot detector that used a modified probabilistic Latent Semantic Analysis (pLSA). Audio features as well as visual concepts such as motion, flame, explosion, and blood were employed. Final integration was achieved though a co-training scheme, typically used when dealing with small amounts of training data and large amounts of unlabeled data. Experimental validation was conducted on 5 movies showing an average F-measure of 0.88.

Most of the approaches are naturally multimodal, exploiting both the image and sound tracks. However, a few works approached the problem based on a single modality. For example, [6] used Gaussian mixture models (GMM) and hidden Markov models (HMM) to model audio events over time series. They considered the presence of gunplay and car racing with audio events such as "gunshot," "explosion," "engine," "helicopter flying," "car braking," and "cheers." Validation was performed on a very restrained data set, containing excerpts of 5 min extracted from 5 movies, leading to an average F-measure of up to 0.90. In contrast, [4] used only visual concepts such as face, blood, and motion information to determine whether an action scene had violent content or not. The specificity of their approach is in addressing more semantics-bearing scene structures of video rather than simple shots.

In general, most of the existing approaches focus more or less on finding the correct concepts that can be translated into violence in general and their findings are bounded by the size of the dataset and the definition of violence. Because of the high variability of violent events in movies, no common and objective enough definition for violent events was ever proposed to the community, even when restricting to physical violence. On the contrary, each piece of work dealing with the detection of violent scenes provides its own definition of the violent events to detect. For instance, [4] targeted "a series of human actions accompanied with bleeding," [11, 32] looked for "scenes containing fights, regardless of context and number of people involved." In [14], the following definition is used: "behavior by persons against persons that intentionally threatens, attempts, or actually inflicts physical harm." In [17], authors were interested in "fast paced scenes which contain explosions, gunshots and person-on-person fighting." Moreover, violent scenes and action scenes are often mixed up in the past as in [5, 17].

The lack of a common definition and the resulting absence of a reference and substantial dataset has made it so far very difficult to compare methods which were sometimes developed for a very specific type of violence. This is precisely the fault that we attempt to correct with the MediaEval violent scene detection task, by creating a benchmark based on a clear and generalizable definition of violence to advance the state of the art on this topic.

8.3 Affect Task Description

The 2011 and 2012 Affect Task required participants to deploy multimodal approaches to automatically detect portions of movies depicting violence. Though not a strict requirement, we tried to emphasize multimodality for several reasons. First, videos are multimodal. Second, violence might be present in all modalities though not necessarily at the same time. This is clearly the case for images and soundtracks. Violence might also be reflected in subtitles though verbal violence was not considered. In spite of a definition of violence limited to physical violence, single modality approaches were bound to be suboptimal and most participants ended up using visual and audio features.

The key for creating a corpus for comparative evaluation clearly remains a general definition of the notion of violence which eases annotation while encompassing a large variety of situations. We discuss here the notion of violence and justify the definition that was adopted before describing the data set and evaluation rules.

8.3.1 Toward a Definition of Violence

The notion of violence remains highly subjective as it depends on viewers. The World Health Organization (WHO) [39] defines violence as: *The intentional use of physical force or power, threatened or actual, against oneself, another person, or against a group or community that either results in or has a high likelihood of resulting in injury, death, psychological harm, maldevelopment, or deprivation.* According to the WHO, three types of violence can be distinguished, namely, self-inflicted, interpersonal, and collective [24]. Each category is divided according to characteristics related to the setting and nature of violence, e.g., physical, sexual, psychological, and deprivation or neglect.

In the context of movies and television, Kriegel [23] defines violence on TV as an *unregulated force that affects the physical or psychological integrity to challenge the humanity of an individual with the purpose of domination or destruction.*

These definitions only focus on intentional actions and, as such, do not include accidents, which are of interest in the use case considered, as they also result in potentially shocking gory and graphic scenes, e.g., a bloody crash. We therefore adopted an extended definition of violence that includes accidents while being as objective as possible and reducing the complexity of the annotation task. In MediaEval 2011 and 2012, violence is defined as *physical violence or accident resulting in human injury or pain.* Violent events are therefore limited to physical violence, verbal, or psychological violence being intentionally excluded.

Although we attempted to narrow the field of violent events down to a set of events as objectively violent as possible, there are still some borderline cases. First of all, sticking to this definition leads to the rejection of some shots in which the results of some physical violence are shown but not the violent act itself. For example, shots in which one can see a dead body with a lot of injuries and blood were not annotated as violent. On the contrary, a character simply slapping another one in the face is considered as a violent action according to the task definition. Other events defined as "intent to kill," in which one sees somebody shooting somebody else for example with the clear intent to kill, but the targeted person escapes with no injury, were also discussed and finally not kept in the violent set. On the contrary, scenes where the shooter is not visible but where shooting at someone is obvious from the audio, e.g., one can hear the gunshot possibly with screams afterward, were annotated as violent. Interestingly, such scenes emphasize the multimodal characteristic of the task. Shots showing actions resulting in pain but with no intent to be violent or, on the contrary, with the aim of helping rather than harming, e.g., segments showing surgery without anesthetics, fit into the definition and were therefore deemed violent.

Another borderline case keenly discussed was the events such as shots showing the destruction of a whole city or the explosion of a moving tank. Technically speaking, these shots do not show any proof of people death or injury, though one can reasonably assume that the city or the tank were not empty at the time of destruction. Consequently, such cases, where pain or injury is implicit, were annotated as violent. Finally, shots showing the violent action and the result of the action itself happen to be separated by several nonviolent shots. In this case, the entire segment was annotated as violent if the duration between the two violent shots (action and result) was short enough (less than 2 s).

8.3.2 Data Description

In line with the use case considered, the dataset consisted of Hollywood movies from a comprehensive range of genres, from extremely violent to movies without violence. In 2011, 15 movies were considered and completed by 3 additional movies in 2012. From these 18 movies, 12 were designated as development data[4] in 2011. The three movies used as test set[5] in 2011 where shifted to the development set in 2012 where three additional movies were provided for evaluation. The list of movies, along with some characteristics, is given in Table 8.1.

The development dataset represents a total of 26,108 shots in 2012—as given by automatic shot segmentation—for a total duration of 102,851 s. Violent content corresponds to 9.25 % of the total duration and 12.27 % of the shots, highlighting the fact that violent segments are not so scarce in this database. We tried to respect the genre distribution (from extremely violent to nonviolent) both in the development and test sets. This appears in the statistics, as some movies such as *Billy Elliot* or *The Wizard of Oz* contain a small proportion of violent shots (around 5 %). The choice we made for the definition of violence impacts the proportion of annotated violence in some movies such as *The Sixth Sense* where violent shots amount to

---

[4] The development data is intended for designing and training the approaches.

[5] The test set data is intended for the official benckmarking.

Table 8.1 Movie dataset (2011 dev. set: first 12 movies; 2011 test set: following 3 movies. 2012 dev. set: first 15 movies; 2012 test set: last three movies)

2012	2011	Movie	Dur	Sh	V-Dur	V-Sh
Dev. set	Dev. set	Armageddon	8680.16	3562	10.16	11.0
		Billy Elliot	6349.44	1236	5.14	4.21
		Eragon	5985.44	1663	11.02	16.6
		Harry Potter 5	7953.52	1891	9.73	12.69
		I am Legend	5779.92	1547	12.45	19.78
		Leon	6344.56	1547	4.3	7.24
		Midnight Express	6961.04	1677	7.28	11.15
		Pirates Carib. 1	8239.4	2534	11.3	12.47
		Reservoir Dogs	5712.96	856	11.55	12.38
		Saving Private Ryan	9751.0	2494	12.92	18.81
		The Sixth Sense	6178.04	963	1.34	2.80
		The Wicker Man	5870.44	1638	8.36	6.72
		Total	**83805.9**	**21608**	**9.02**	**14.8**
	Test set	Kill Bill	6370.4	1597	17.47	23.98
		The Bourne Identity	6816.0	1995	7.61	9.22
		The Wizard of Oz	5859.2	908	5.51	5.06
		Total	**19045.6**	**4500**	**11.55**	**13.62**
	Total		**102851.5**	**26108**	**9.25**	**12.27**
Test set		Dead Poets Society	7413.2	1583	1.5	2.14
		Fight Club	8005.7	2335	13.51	13.27
		Independance Day	8834.3	2652	9.92	13.98
	Total		**24253.2**	**6570**	**8.53**	**10.88**

Dur duration in seconds; *Sh* number of shots; *V-Dur* violent shot duration proportion (%); *V-Sh* Violent shot proportion (%)

only 2.8 % of the duration. However, the movie contains several shocking scenes of dead people which do not fit the definition of violence that we adopted. In a similar manner, psychological violence, such as what may be found in *Billy Elliot*, was also not annotated, which also explains the small number of violent shots in this particular movie.

The violent scenes dataset was created by seven human assessors. In addition to segments containing physical violence according to the definition adopted, annotations also include high-level concepts potentially related to violence for the visual and audio modalities, highlighting the multimodal character of the task.

The annotation of violent segments was conducted using a 3 step process, with the same so-called "master annotators" for all movies. A first master annotator extracted all violent segments. A second master annotator reviewed the annotated segments and possibly missed segments according to his/her own judgment. Disagreements were discussed on a case by case basis, the third master annotator making the final decision in case of an unresolved disagreement. Each annotated violent segment contained a single action, whenever possible. In the case of overlapping actions, the corresponding global segment was proposed as a whole. This was indicated in the annotation files by adding the tag "multiple action scene." The boundaries of each violent segment were defined at the frame level, i.e., indicating the start and end frame numbers.

The high-level video concepts were annotated through a simpler process, involving only two annotators. Each movie was first processed by an annotator and then reviewed by one of the master annotators.

Seven visual concepts are provided: *presence of blood, fights, presence of fire, presence of guns, presence of cold weapons, car chases and gory scenes*. For the benchmark, participants had the option to carry out detection of the high-level concepts. However, concept detection is not among the task's goals and these high-level concept annotations were only provided on the development set. Each of these high-level concepts followed the same annotation format as for violent segments, i.e., starting and ending frame numbers and possibly some additional tags which provide further details. For blood annotations, a tag in each segment specifies the proportion of the screen covered in blood. Four tags were considered for fights: only two people fighting, a small group of people (roughly less than 10), large group of people (more than 10), distant attack (i.e., no real fight but somebody is shot or attacked at distance). As for the presence of fire, anything from big fires and explosions to fire coming out of a gun while shooting, a candle, a cigarette lighter, a cigarette, or sparks was annotated, e.g., a space shuttle taking off also generates fire and receives a fire label. An additional tag may indicate special colors of the fire (i.e., not yellow or orange). If a segment of video showed the presence of firearms (respectively cold weapons) it was annotated by any type of (parts of) guns (respectively cold weapons) or assimilated arms. Annotations of gory scenes are more difficult. In the present task, they are indicating graphic images of bloodletting and/or tissue damage. It includes horror or war representations. As this is also a subjective and difficult notion to define, some additional segments showing disgusting mutants or creatures are annotated as gore. In this case, additional tags describing the event/scene are added.

For the audio modality, three audio concepts were annotated, namely, *gunshots, explosions, screams*. Those concepts were extracted using the English audio tracks. Contrary to what is done for the video concepts, audio segments are identified by start and end times in seconds. Additional tags may be added to each segment to distinguish different types of subconcepts. For instance, distinction was made between gunshots and cannon fires. All kinds of explosions were annotated, even magic explosions as well as explosions resulting from shells or cannonballs in cannon fires. Last, scream annotations are also provided, however for 9 movies only, in which anything from nonverbal screams to what was called "effort noise" was extracted, as long as the noise came from a human or a humanoid. Effort noises were separated from the rest, by the use of two different tags in the annotation.

In addition to the annotation data, automatically generated shot boundaries with their corresponding key frames, as detected by Technicolor's software, were also provided with each movie.

8.3.3 Evaluation Rules

Due to copyright issues, the video content was not distributed and participants were required to buy the DVDs. Participants were allowed to use all information automatically extracted from the DVDs, including visual and auditory material as well as subtitles. English was the chosen language for both the audio and subtitles channels. The use of any other data, not included in the DVD (web sites, synopsis, etc.) was not allowed.

Two types of runs were initially considered in the task, a mandatory shot classification run and an optional segment detection one. The shot classification run consisted in classifying each shot provided by Technicolor's shot segmentation software as violent or not. Decisions were to be accompanied by a confidence score where the higher the score, the more likely the violence. Confidence scores were optional in 2011 and compulsory in 2012 because of the chosen metric. The segment detection run involved detection of the violent segment boundaries, regardless of the shot segmentation provided.

System comparison was based on different metrics in 2011 and 2012. In 2011, performance was measured using a detection cost function weighting false alarms (FA) and missed detections (MI), according to

$$C = C_{\text{fa}} \cdot P_{\text{fa}} + C_{\text{miss}} \cdot P_{\text{miss}} \tag{8.1}$$

where the costs $C_{\text{fa}} = 1$ and $C_{\text{miss}} = 10$ were arbitrarily defined to reflect (a) the prior probability of the situation and (b) the cost of making an error. P_{fa} and P_{miss} are the estimated probabilities of respectively false alarms (false positive) and missed detections (false negative) given the system's output and the reference annotation. In the shot classification, the FA and MI probabilities were calculated on a per shot basis while in the segment level run, they were computed on a per unit of time basis, i.e., durations of both references and detected segments are compared. This cost function is called "MediaEval cost" in all that follows.

Experience taught us that the MediaEval detection cost was too strongly biased toward low-missed detection rates, leading to systems hardly reaching cost values lower than 1 and therefore worse than a naive system classifying all shots as violent. We therefore adopted the Mean Average Precision (MAP) computed over the first 100 top-ranked violent segments as evaluation metric. Note that this measure is also well adapted to the search-related use case that serves as a basis for our work.

We also report detection error tradeoff curves, showing P_{fa} as a function of P_{miss} given a segmentation and the confidence score for each segment, to compare potential performance at different operating points. Note that in the segment detection run, DET curves are possible only for systems returning a dense segmentation (a list of segments that spans the entire video): segments not present in the output list are considered as non violent for all thresholds.

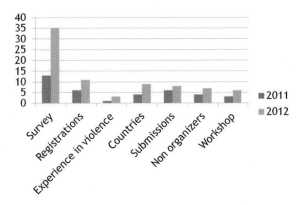

Fig. 8.1 Evolution of the participation to the task between 2011 and 2012

8.4 Results

In 2011, the Affect Task on Violent Scenes Detection was proposed in MediaEval as a pilot for the first year. Thirteen teams, corresponding to 16 research groups considering joint submission proposals, declared interest in the task. Finally, six teams registered and completed the task, representing four different countries, for a grand total of 29 runs submitted. These figures show the interest for the task for this first year. This was confirmed in 2012, with the registration of 11 teams, of which 8 crossed the final line, by sending 36 runs for the evaluation. Interest is also emphasized by the wide geographic coverage area of teams. Interestingly, the multimodal aspect of the task shows in the fact that participants come from different communities, namely the audio and image processing communities. A more detailed evolution of the task for these two years is summarized in Fig. 8.1.

Official results are reported in Table 8.2. Despite the change of official metric between 2011 and 2012, MAP values were also computed on the 2011 submissions. Similarly, the MediaEval cost is reported for 2012. It should nevertheless be noted that these two metrics imply different tunings of the systems (toward low precision rate for the MediaEval cost, and on the contrary toward high precision for the MAP), meaning that metric values should be compared cautiously, as systems were not optimized in the same way.

In 2011 and 2012, all participants submitted predominantly runs for the shot classification task. Only the ARF team submitted one segment level run in 2012. Results show a substantial improvement between 2011 and 2012. Although the overall performances of the proposed systems in 2011 were not good enough to satisfy the requirements of a real-life commercial system, in 2012 three systems reached MAP@100 values above 60%, leading to the conclusion that research still needs to be conducted on this subject, nevertheless state-of-the-art systems already show convincing performances.

Detection error trade-off curves, obtained from the confidence values provided by participants, are given in Fig. 8.2 for the best run of each participant according

Table 8.2 Official results of the 2011 and 2012 Affect task evaluation at MediaEval

Team	Country	MAP@20	MAP@100	Med. cost
			2011 benchmark	
ARF	Austria-Romania-France	–	–	–
DYNI	France	13.81 (*31.22*)	18.33 (*19.07*)	6.46 (*7.57*)
LIG	France	23.87 (*23.87*)	18.01 (*18.01*)	7.93 (*7.93*)
NII	Japan	40.73 (*33.14*)	24.78 (*27.71*)	1 (*1*)
Shanghai-Hongkong	China	–	–	–
TEC*	France-UK	33.33 (*44.94*)	21.89 (*40.58*)	0.76 (*0.89*)
TUB	Germany	4.69 (*4.69*)	14.29 (*14.29*)	1.26 (*1.26*)
TUM	Germany-Austria	–	–	–
UNIGE*	Switzerland	29.28 (*29.28*)	24.57 (*24.57*)	2.00 (*2.83*)
			2012 benchmark	
ARF	Austria-Romania-France	70.08	65.05	3.56
DYNI	France	0	12.44	7.96
LIG	France	28.64	31.37	4.16
NII	Japan	40.07	30.82	1.28
Shanghai-Hongkong	China	73.6	62.38	5.52
TEC*	France-UK	66.89	61.82	3.56
TUB	Germany	35.92	18.53	4.2
TUM	Germany-Austria	50.42	48.43	7.83
UNIGE*	Switzerland	–	–	–

In 2011, we report in plain figure results from the best run according to the MediaEval cost and indicate in parenthesis results corresponding to the best run according to the mean average precision. Team names indicated with "*" correspond to the task organizers

to the official metric for the year considered. Clearly, ordering of the systems differs according to the operating point. Once again the direct comparison of the 2011 and 2012 curves is to be considered with caution. Nevertheless, improvements can be observed between the 2 years. Whereas in 2011, only one participant reached at best a false alarm rate of 20% for a missed detection rate of about 25%, in 2012, at least two participants have similar results and three more additional teams have fair results.

Analyzing the 2011 submissions, three different systems categories can be distinguished. Two participants (NII [26] and LIG [37]) treated the problem of violent scene detection as a concept detection problem, applying generic systems developed for TRECVid evaluations to violent scene detection, potentially with specific tuning. Both sites used classic video only features, computed on the key frames provided, based on color, textures, edges, either local (interest points) or global, and classic classifiers. One participant (DYNI [15]) proposed a classifier-free technique exploiting only two low-level audio and video features, computed on each successive frame, both measuring the activity within a shot. After a late fusion process, decisions were taken by comparison with a threshold. The last group of participants (TUB [2], UGE [16] and TI [33]) built dedicated supervised classification systems for the task of violent scene detection. Different classifiers were used from SVM, Bayesian

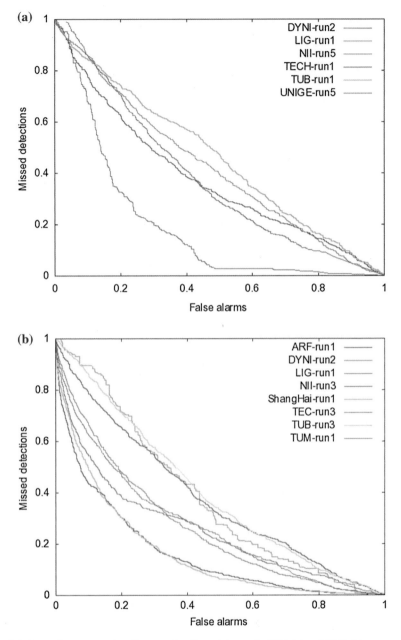

Fig. 8.2 Detection error trade-off *curves* for all participants in 2011 (**a**) and 2012 (**b**)

networks to linear or quadratic discriminant analysis. All used multimodal features, either audio-video or audio-video-textual features (UGE). Features were computed globally for each shot (UGE, TI) or on the provided key frames (TUB).

In 2012, systems were all supervised classification systems; LIG [10] and NII [25] went on with some improved versions of their generic systems dedicated to concept detection, while others implemented dedicated versions of such systems for the task of violent scene detection. Chosen classifiers were mostly SVM, with some exceptions for neural networks and Bayesian networks. It should be noted that most participants [1, 10, 13, 22, 35, 38] voted for multimodal (audio + video) systems and that multimodality seems to help the performance of such systems. Globally, classic low-level audio (MFCC, zero-crossing rate, asymetry, roll-off, etc.) and video (color histograms, texture-related, Scale Invariant Feature Transform-like, Histograms of Oriented Gradients, visual activity, etc.) features were extracted. One exception may be noted with the use of multi-scale local binary pattern histogram features by DYNI [30]. Added to those classical features, audio and video mid-concept detection was also used for this second year [10, 22, 25, 38], thanks to the annotated high-level concepts. Such mid-level concepts, especially used in a two-step classification scheme [38], seem to be promising.

Based on these results, one may draw some tentative conclusions about the global characteristics that were more likely to be useful for violence detection. Local video features (SIFT-like) did not add a lot of information to the systems. On the contrary, taking advantage of different modalities seems to improve performance, especially when modalities are merged using late fusion. Although results do not prove their impact in one way or another, it also seems of interest to use temporal integration. This was carried out in different manners in the systems, either by using contextual features, i.e., features at different times, or by temporal smoothing or aggregation of the decisions at the output of the chain. Using intermediate concept detection with high-level concepts related to violence such as those provided in the task seems to be rewarding.

8.5 Multimodal Approaches

Progress achieved between 2011 and 2012 can probably be explained by two main factors. Data availability is undoubtedly the first one, along with experience on the task. Exploiting multimodal features is also one of the keys. While many systems made very limited use of multiple modalities in 2011, multimodal integration became more widely spread, mostly relying on the audio and visual modalities.

We provide here details for two multimodal systems which competed in 2012, namely the ARF system based on mid-level concepts detected from multimodal input and the Technicolor/IRISA system which directly exploits a set of low-level audio and visual features.

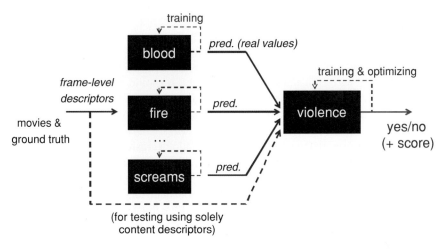

Fig. 8.3 Description of ARF teams's system developed for MediaEval 2012 (*black boxes* refer to classifiers)

8.5.1 A Mid-Level Concept Fusion Approach

We describe the approach developed by the ARF team [21, 38], relying on fusing mid-level concept predictions inferred from low-level features by employing a bank of multilayer perceptron classifiers featuring a dropout training scheme.

The motivation of this approach lies in the high variability in appearance of violent scenes in movies and the low amount of training data that is usually available. In this scenario, training a classifier to predict violent frames directly from visual and auditory features seems rather difficult. The system proposed by ARF team uses the task provided a high-level concept ground-truth to infer mid-level concepts as an intermediate step toward the final violence detection goal, thus attempting to limit the semantic gap. Experiments proved that predicting mid-level concepts from low-level features should be more feasible than directly predicting all forms of violence.

8.5.1.1 Description of the System

Violence detection is first carried out at frame level by classifying each frame as being violent or nonviolent. Segment level prediction (shot level or arbitrary length) is then determined by a simple aggregation of frame level decisions. Given the complexity of this task, i.e., labeling of individual frames rather than video segments (ca. 160,000 frames per movie), the classification is tackled by exploiting the inherent parallel architecture of neural networks. The system involves several processing steps as illustrated in Fig. 8.3.

Multimodal features: First, raw video data is converted into content descriptors whose objective is to capture meaningful properties of the auditory-visual information. Feature extraction is carried out at the frame level. Given the specificity of the task, the system was tested using audio, color, feature description, and temporal structure information, which is specific both for violence-related concepts as well as for the violent content itself. Results reported in 2012 were obtained with the following descriptors:

- audio descriptors (196 dimensions) consist of general purpose descriptors: linear prediction coefficients, line spectrum pairs, MFCCs, zero-crossing rate, and spectral centroid, flux, rolloff, and kurtosis, augmented with the variance of each feature over a window of 0.8 s around the current frame[6];
- color descriptors (11 dimensions) using the color naming histogram proposed in [12] which maps colors to 11 universal color names ("black", "blue","brown", "gray", "green", "orange", "pink", "purple", "red", "white", and "yellow");
- visual features (81 dimensions) which consist of the 81-dimensional Histogram of Oriented Gradients [29];
- temporal structure (1 dimension) derives a measure of visual activity. The cut detector in [20] that measures visual discontinuity by means of a difference between color histograms of consecutive frames, was modified to account for a broader range of significant visual changes. For each frame it determines the number of detections in a certain time window centered at the current frame. High values of this measure will account for important visual changes that are typically related to action.

Neural network classification: Both at the concept level and at the violence level, classification is carried out with a neural network, namely a multilayer perceptron with a single hidden layer of 512 logistic sigmoid units. Network is trained by gradient descent on the cross-entropy error with backpropagation [36], using the recent idea in [19] to improve generalization: For each presented training case, a fraction of input and hidden units is omitted from the network and the remaining weights are scaled up to compensate. The set of dropped units is chosen at random for each presentation of a training case, such that many different combinations of units will be trained during an epoch.

Concept detection consists of a bank of perceptrons that are trained to respond to each of the targeted violence-related concepts, such as presence of "fire," presence of "gunshots," or "gory" scenes (see Sect. 8.3.2). As a result, a concept prediction value in [0, 1] is obtained for each concept. These values are used as inputs to a second classifier, acting as a final fusion scheme to provide values for the two classes "violence" and "nonviolence" on a frame-by-frame basis. For all classifiers, parameters were trained using reference annotations coming along with the data.

Violence classification: Frame prediction of violence for the unlabeled data is given by the system's output when fed with the new data descriptors. As prediction is provided at frame level, aggregation into segments is performed by assigning a violence score corresponding to the highest predictor output for any frame within the segment. The segments are then tagged as "violent" or "nonviolent" depending on whether

---

[6] The Yaafe toolkit for audio feature extraction was used.

Table 8.3 ARF team violence shot-level detection results at MediaEval 2012

Run	Modality	Precision (%)	Recall (%)	F_1-score (%)
ARF-(c)	Concepts	46.14	54.40	49.94
ARF-(a)	Audio	46.97	45.59	46.27
ARF-(av)	Audio-visual	32.81	67.69	44.58
ARF-(avc)	Audio-visual	31.24	66.15	42.44
ARF-(v)	Visual	25.04	61.95	35.67

their violence score exceeds a certain threshold (determined in the training step of the violence classifier).

8.5.1.2 Results

Results are evaluated on the shot classification task and on the segment detection one.

Shot level classification: To highlight the contributions of the concept fusion scheme, different feature combinations were tested, namely: ARF-(c) uses as features only mid-level concept predictions for violence detection; ARF-(a) uses only audio descriptors, i.e., the violence classifier is trained directly on features instead of using the concept prediction outputs; ARF-(v) uses only visual features; ARF-(av) uses only audio-visual features; finally, ARF-(avc) uses all concepts and audio-visual features using an early fusion aggregation of concept predictions and features.

Results on the 2012 benchmark, reported in Table 8.3, exhibited a F-measure of 49.9 which placed the system among the top systems. The lowest discriminative power is achieved using only visual descriptors (ARF-(v)), with an F-measure of 35.6. Compared to visual features, audio features seem to show better descriptive power, providing an F-measure of 46.3. The combination of descriptors (early fusion) tends to reduce their efficiency and yields lower performance than the use of concepts alone, e.g., audio-visual (ARF-(av)) yields an F-measure of 44.6, while audio-visual-concepts (ARF-(avc)) achieve 42.4.

Figure 8.4 details the precision-recall curves for this system. The use of concepts fusion scheme (red line) proved again to provide significantly higher recall than the sole use of audio-visual features or the combination of all for a precision of 25 % and above.

Arbitrary segment-level results: At the segment detection level, the use of the fusion of the mid-level concepts achieves average precision and recall values of 42.21 and 40.38 %, respectively, while the F-measure is 41.3. This yields a miss rate (at time level) of 50.69 % and a very low false alarm rate of only 6 %. These results are promising considering the difficulty of precisely detecting the exact time interval of violent scenes, but also the subjectivity of the human assessment (reflected in the ground truth).

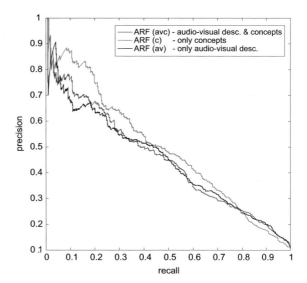

Fig. 8.4 ARF system precision-recall *curves* [21]

8.5.2 Direct Modeling of Multimodal Features

We describe here the approach adopted in the joint submission of Technicolor and IRISA in 2012, which directly models a set of multimodal features to infer violence at the shot level. Relying on Bayesian networks and, more specifically, on structure learning in Bayesian networks [18], we investigate multimodal integration via early and late fusion strategies, together with temporal integration.

8.5.2.1 Description of the System

Figure 8.5 provides a schematic overview of the various steps implemented in Technicolor's system. Violence detection is performed at the shot level via direct modeling of audio and visual features aggregated over shots. Classification is then performed either based on the entire set of multimodal features or independently for each modality. In this last case, late fusion is used to combine modalities. In both cases, temporal information can be used at two distinct levels: in the model with contextual features or as a postprocessing step to smooth decisions taken on a per shot basis.

Multimodal features: For each shot, different low-level features are extracted from both the audio and the video signals of the movies:

- Audio features: the audio features, extracted using 40 ms frames with 20 ms overlap, are: the energy (E), the frequency centroid (C), the asymmetry (A), the flatness (F), the 90 % frequency roll-off (R), and the zero-crossing rate (Z) of the signal. These features are normalized to zero mean and unit variance, and averaged over

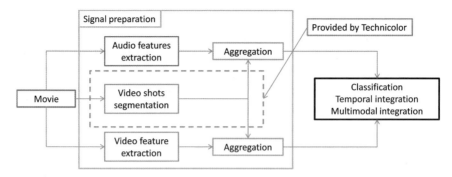

Fig. 8.5 Description of the technicolor/IRISA system at MediaEval 2012

the duration of a shot, in order to obtain a single value per shot for each feature. The audio feature vector dimension is $D = 6$;
- Video features: the video features extracted per shot are: the shot length (SL), the mean proportion of blood color pixels (B), the mean activity (AC), the number of flashes (FL), the mean proportion of fire color pixels (FI), a measurement of color coherence (CC), the average luminance (AVL), and three color harmony features, the majority harmony template (Tp), the majority harmony template mean angle (Al), and the majority harmony template mean energy (Em) [3]. The feature vector dimension is $D = 10$.

Features are quantized in 21 bins on a per movie basis, except for the majority template whose values are already quantized over 9 bins.

Bayesian network classification: Bayesian networks are used as a classification technique. The idea behind Bayesian networks is to build a probabilistic network on top of the input features with a node in the network for classification of violence. The network represents conditional dependencies and independencies between the features, and it is possible to learn the structure of the graph using structure learning algorithms. The output of the classifier is, for each shot, the estimated posterior probabilities for each class, viz., violence and nonviolence.

We compared a so-called naive structure, which basically links all the features to the class variable, with structures learned using either forest-augmented networks (FAN) [28] or K2 [7]. The FAN structure consists in building a tree on top of the naive structure based on some criterion related to classification accuracy. On the contrary, the K2 algorithm does not impose the naive structure but rather attempts a better description of the data based on a Bayesian information criterion, thus not necessarily targeting better classification.

Temporal integration: Two strategies for integrating temporal information were tested. The first one is a contextual representation of the shots at the input of the classifier, where classification of a shot relies on the features for this shot augmented with the features from the neighboring shots. If we denote F_i the features for shot i, the contextual representation of shot i is given by:

$$F_i^\star := \{F_{i-n}, F_{i-n+1}, \ldots, F_{i-1}, F_i, F_{i+1}, \ldots, F_{i+n-1}, F_{i+n}\} \quad (8.2)$$

where the context size was set to $n = 5$ (empirically determined).

In addition to contextual representation, we also used temporal filtering to smooth the shot by shot independent classification, considering two types of filters:

- a majority vote over a sample window of size $k = 5$, after thresholding the probabilities.
- an average of the probabilities over a sliding window of size $k = 5$, before thresholding the probabilities.

Contrary to averaging, majority vote does not directly provide a confidence score in the decision taken. We implemented the following heuristics in this case. For a given shot, if the vote results in violence, the confidence score is set to $\min\{P(S_v)\}$, where $P(S_v)$ is the set of probabilities of the shots that were considered as violent within the window. If the vote results in a nonviolent decision, the confidence score is set to $\max\{P(S_{nv})\}$, where $P(S_{nv})$ is the set of probabilities of the shots that were considered as nonviolent within the window.

Multimodal integration: As for multimodal integration, early fusion and late fusion are compared. Early fusion consists in the concatenation of the audio and the video attributes in a common feature vector. The violence classifier is then learned using this feature vector. Late fusion consists in fusing the outputs of both a video classifier and an audio classifier. In order to fuse the outputs of the ith shot, the following rule is used:

$$P_{\text{fused}}^{S_i}(P_{v_a}^{S_i}, P_{v_v}^{S_i}) = \begin{cases} \max\{P_{v_a}^{S_i}, P_{v_v}^{S_i}\} & \text{if both decisions are violent} \\ \min\{P_{v_a}^{S_i}, P_{v_v}^{S_i}\} & \text{if both decisions are nonviolent} \\ P_{v_a}^{S_i} \cdot P_{v_v}^{S_i} & \text{otherwise} \end{cases} \quad (8.3)$$

where $P_{v_a}^{S_i}$ (resp. $P_{v_v}^{S_i}$) is the probability that shot i is violent as given by the audio (respectively video) classifier. This simple rule of thumb yields a high score when both classifiers agree on violence, and a low score when they agree on nonviolent.

8.5.2.2 Results

We first compare the different strategies implemented using cross-validation over the 15 development movies, leaving one movie out for test on each fold. We then report results for the best configuration on the official 2012 evaluation.

The MAP@100 values obtained in cross-validation for the audio only, the video only, and the early fusion experiments are presented in Table 8.4. For the late fusion experiments, all classifier combinations, i.e., the naive structure, the FAN, or the K2 networks, with or without context, with or without temporal filtering, have been tested. The seven best combinations are presented in Table 8.5.

It is interesting to note that, while the FAN networks are supposed to perform well in classification, they are outclassed by the K2 and the naive structures in these exper-

Table 8.4 MAP@100 values obtained via cross-validation

Network structure	Context	Audio			Video			Early fusion		
		1	2	3	1	2	3	1	2	3
Naive	No	36.3	**39.4**	38.4	25.4	**30.0**	27.9	36.0	**40.3**	37.5
	Yes	36.9	36.2	**37.3**	31.1	30.8	**31.3**	38.5	37.1	**38.5**
FAN	No	26.9	**30.9**	29.3	22.4	**26.9**	25.0	29.0	34.7	**34.8**
	Yes	20.1	20.6	**21.4**	25.5	**27.4**	26.9	25.6	**26.2**	26.1
K2	No	36.3	**39.1**	37.8	26.0	**30.7**	29.0	37.4	**40.9**	39.2
	Yes	36.1	**39.0**	37.0	27.0	27.5	**27.9**	32.3	32.3	**33.2**

Results are reported for the audio and the video modalities, and for early fusion. For each modality, column 1 corresponds to no temporal filter, column 2 to a sliding window averaging, and column 3 to a majority vote

Table 8.5 Results obtained for the seven best late fusion parameter combinations

S_a	C_a	S_v	C_v	T_c	T_{lf}	MAP@100
K2	No	Naive	Yes	1	2	43.18
K2	Yes	Naive	Yes	3	2	42.59
K2	Yes	Naive	Yes	1	2	42.55
K2	Yes	Naive	Yes	2	2	42.53
Naive	No	Naive	Yes	3	2	42.45
K2	No	Naive	Yes	3	3	42.36
Naive	No	Naive	Yes	3	3	42.32

S_a Audio structure, C_a Audio context, S_v Video structure, C_v Video context, T_c Temporal filter applied to the classifiers, T_{lf} Temporal filter applied after late fusion

iments. As for the other two types of structure, they both seem to provide equivalent results, which shows that structure learning is not always beneficial. One must also note that, if the influence of context is not always clear for the modalities presented in Table 8.4, temporal filters systematically improve the results, thus showing the importance of the temporal aspect of the signal. However, it is not possible to say which filter provides the best performances. Finally, the importance of multimodal integration is clearly shown as the best results were obtained via both early and late fusions. The importance of temporal integration is further reinforced by the results obtained via late fusion: among the best combinations, the contextual naive structure is always used for the video modality, and a temporal filter is always used after the fusion step. Moreover, it seems that late fusion performs better than early fusion.

The system chosen and submitted to the 2012 campaign is the best system obtained via late fusion. This system uses a noncontextual K2 network for the audio modality, a contextual naive network for the video modality, and a sliding window probability averaging filter after the fusion. It is applied to the test movies and the obtained results are presented in Table 8.6.

The first thing to note is that results are much better than in the cross-validation experiments ($\simeq +18\,\%$). Taking a closer look at the individual results for each movie, it appears that the lowest results are obtained for the movie *Fight Club*, while for

Table 8.6 Results obtained on the test movies

Movie	P	R	F1	MAP@100	MC
Dead poet society	5.06	64.71	9.38	60.56	4.09
Fight club	25.14	58.06	35.09	53.15	3.70
Independence day	26.22	75.20	38.89	71.76	1.35
Total	21.72	67.27	32.83	61.82	3.57

Column *P* corresponds to the Precision, *R* to the Recall, F1 of the F1-measure, and MC to the MediaEval Cost. The values in the MAP@100 column presented for each movie actually correspond to the average precision over the first hundred top-ranked samples (AP@100), the MAP@100 being the value in the Total row

the other systems presented in the 2012 campaign, the lowest results were usually obtained for *Dead Poet Society*. This is encouraging as, contrary to the other systems, this system was able to cope with such a nonviolent movie. The "low" results obtained for *Fight Club* can be explained by the very particular type of violence present in this movie, which might be under-represented in the training database. Similarly, the good results obtained for *Independence day* can be explained by its similarity with the movie *Armageddon* present in the training set.

These results clearly emphasize again the importance of multimodal integration, through late fusion of classifiers. Finally, the overall result of 61.82 for the MAP@100 is already convincing for the evolution of the task towards real-life commercial systems.

8.6 Conclusions

Running the Violent Scene Detection task in the framework of the MediaEval benchmark initiative for 2 years have resulted in two major results: a comprehensive data set to study violence detection in videos, with a focus on Hollywood movies; state-of-the-art multimodal methods which establish a baseline for future research to compare with. Results in the evaluation, demonstrated by the two systems described in this chapter, clearly emphasize the crucial role of multimodal integration, either for mid-level concept detection or for direct detection of violence. The two models compared here, namely Bayesian networks and neural networks, have proven beneficial to learn relations between audio and video features for the task of violence detection.

Many questions are still to be addressed, among which we believe two to be crucial. First, Bayesian networks with structure learning, as well as neural networks, implicitly learn the relations between features for better classification. Still, it was observed that late fusion performs similarly. There is therefore a need for better models of the multimodal relations. Second, mid-level concept detection has proven beneficial, reducing the semantic gap between features and classes of interest. There is however, still a huge gap between features and concepts such as gunshots, screams, or explosions, as demonstrated by various experiments [21, 34]. An interesting idea for the future is that of inferring concepts in a data-driven manner, letting the data define concepts whose semantic interpretation is to be found post-hoc. Again, Bayesian networks and neural networks might be exploited to this end, with hidden nodes whose meaning have to be inferred.

Acknowledgments This work was partially supported by the Quaero Program. We would also like to acknowledge the MediaEval Multimedia Benchmark for providing the framework to evaluate the task of violent scene detection. We also greatly appreciate our participants for giving us their consent to describe their systems and results in this paper. More information about the MediaEval campaign is available at: http://www.multimediaeval.org/. The working note proceedings of the MediaEval 2011 and 2012 which included the participants' contributions can be found online at http://www.ceur-ws.org/Vol-807 and http://www.ceur-ws.org/Vol-927, respectively.

References

1. Acar E, Albayrak S (2012) Dai lab at mediaeval 2012 affect task: the detection of violent scenes using affective features. In: MediaEval 2012, multimedia benchmark workshop
2. Acar E, Spiegel S, Albayrak S (2011) Mediaeval 2011 affect task: Violent scene detection combining audio and visual features with svm. In: MediaEval 2011, multimedia benchmark workshop
3. Baveye Y, Urban F, Chamaret C, Demoulin V, Hellier P (2013) Saliency-guided consistent color harmonization. Computational color imaging, Lecture notes in computer science, vol 7786. Springer, Berlin, pp 105–118
4. Chen LH, Hsu HW, Wang LY, Su CW (2011) Violence detection in movies. In: 8th IEEE international conference on computer graphics, imaging and visualization (CGIV 2011), pp 119–124
5. Chen LH, Su CW, Weng CF, Liao HYM (2009) Action Scene Detection With Support Vector Machines. J Multimedia 4:248–253. doi:10.4304/jmm.4.4.248-253
6. Cheng WH, Chu WT, Wu JL (2003) Semantic context detection based on hierarchical audio models. In: Proceedings of the 5th ACM SIGMM international workshop on multimedia information retrieval, pp 109–115
7. Cooper GF, Herskovits E (1992) A Bayesian method for the induction of probabilistic networks from data. Mach Learn 9:309–347. http://dx.doi.org/10.1007/BF00994110
8. Datta A, Shah M, Da Vitoria Lobo N (2002) Person-on-person violence detection in video data. In: Proceedings of 16th IEEE international conference on pattern recognition, vol 1. pp 433–438
9. Demarty CH, Penet C, Gravier G, Soleymani M (2012) A benchmarking campaign for the multimodal detection of violent scenes in movies. In: Computer Vision-ECCV 2012. Workshops and demonstrations, Springer, pp 416–425
10. Derbas N, Thollard F, Safadi B, Quénot G (2012) Lig at mediaeval 2012 affect task: use of a generic method. In: MediaEval 2012, multimedia benchmark workshop
11. de Souza FDM, Chávez GC, do Valle E, de A Araujo A (2010) Violence detection in video using spatio-temporal features. In: 23rd IEEE conference on graphics, patterns and images (SIBGRAPI 2010), pp 224–230
12. de Weijer JV, Schmid C, Verbeek J, Larlus D (2009) Learning color names for real-world applications. IEEE Trans Image Process 18(7):1512–1523
13. Eyben F, Weninger F, Lehment N, Rigoll G, Schuller B (2012) Violent scenes detection with large, brute-forced acoustic and visual feature sets. In: MediaEval 2012, multimedia benchmark workshop

14. Giannakopoulos T, Makris A, Kosmopoulos D, Perantonis S, Theodoridis S (2010) Audio-visual fusion for detecting violent scenes in videos. In: Konstantopoulos S et al (eds) Artificial intelligence: theories, models and applications,Lecture notes in computer scienc, vol 6040. Springer, pp 91–100
15. Glotin H, Razik J, Paris S, Prevot JM (2011) Real-time entropic unsupervised violent scenes detection in hollywood movies - dyni @ mediaeval affect task 2011. In: MediaEval 2011, multimedia benchmark workshop
16. Gninkoun G, Soleymani M (2011) Automatic violence scenes detection: a multi-modal approach. In: MediaEval 2011, multimedia benchmark workshop
17. Gong Y, Wang W, Jiang S, Huang Q, Gao W (2008) Detecting violent scenes in movies by auditory and visual cues. In: Huang YM et al (eds) Advances in multimedia information processing - (PCM 2008), Lecture notes in computer science, vol 5353. Springer, pp 317–326
18. Gravier G, Demarty CH, Baghdadi S, Gros P (2012) Classification-oriented structure learning in bayesian networks for multimodal event detection in videos. Multimedia tools and applications, pp 1–17. doi:10.1007/s11042-012-1169-y, http://dx.doi.org/10.1007/s11042-012-1169-y
19. Hinton G, Srivastava N, Krizhevsky A, Sutskever I, Salakhutdinov R (2012) Improving neural networks by preventing co-adaptation of feature detectors. http://arxiv.org/abs/1207.0580
20. Ionescu B, Buzuloiu V, Lambert P, Coquin D (2006) Improved cut detection for the segmentation of animation movies. In: IEEE international conference on acoustics, speech, and signal processing
21. Ionescu B, Schlüter J, Mironică I, Schedl M (2013) A naive mid-level concept-based fusion approach to violence detection in hollywood movies. In: Proceedings of the 3rd ACM international conference on multimedia retrieval, pp 215–222
22. Jiang YG, Dai Q, Tan CC, Xue X, Ngo CW (2012) The shanghai-hongkong team at mediaeval2012: Violent scene detection using trajectory-based features. In: MediaEval 2012, multimedia benchmark workshop
23. Kriegel B (2003) La violence à la télévision. rapport de la mission d'évaluation, d'analyse et de propositions relative aux représentations violentes à la télévision. Technical report, Ministère de la Culture et de la Communication, Paris
24. Krug EG, Mercy JA, Dahlberg LL, Zwi AB (2002) The world report on violence and health. The Lancet 360(9339):1083–1088 (2002). doi: 10.1016/S0140-6736(02)11133-0. http://www.sciencedirect.com/science/article/pii/S0140673602111330
25. Lam V, Le DD, Le SP, Satoh S, Duong DA (2012) Nii, Japan at mediaeval 2012 violent scenes detection affect task. In: MediaEval 2011, multimedia benchmark workshop
26. Lam V, Le DD, Satoh S, Duong, DA (2011) Nii, Japan at mediaeval 2011 violent scenes detection task. In: MediaEval 2011, multimedia benchmark workshop
27. Lin J, Wang W (2009) Weakly-supervised violence detection in movies with audio and video based co-training. In: Advances in multimedia information processing-PCM 2009, Springer, pp 930–935
28. Lucas P (2002) Restricted Bayesian network structure learning. In: Advances in Bayesian networks, studies in fuzziness and soft computing, pp 217–232
29. Ludwig O, Delgado D, Goncalves V, Nunes U (2009) Trainable classifier-fusion schemes: An application to pedestrian detection. In: IEEE internation conference on intelligent transportation systems, pp 432–437
30. Martin V, Glotin H, Paris S, Halkias X, Prevot JM (2012) Violence detection in video by large scale multi-scale local binary pattern dynamics. In: MediaEval 2012, multimedia benchmark workshop
31. Nam J, Alghoniemy M, Tewfik AH (1998) Audio-visual content-based violent scene characterization. In: Proceedings of IEEE international conference on image processing (ICIP-98), vol 1. pp 353–357
32. Nievas EB, Suarez OD, García GB, Sukthankar R (2011) Violence detection in video using computer vision techniques. In: Computer analysis of images and patterns, Springer, pp 332–339

33. Penet C, Demarty CH, Gravier G, Gros P (2011) Technicolor and inria/irisa at mediaeval 2011: learning temporal modality integration with bayesian networks. In: MediaEval 2011, Multimedia Benchmark Workshop, CEUR Workshop Proceedings, vol 807. http://CEUR-WS.org
34. Penet C, Demarty CH, Gravier G, Gros P (2013) Audio event detection in movies using multiple audio words and contextual Bayesian networks. In: Workshop on content-based multimedia indexing
35. Penet C, Demarty CH, Soleymani M, Gravier G, Gros P (2012) Technicolor/inria/imperial college london at the mediaeval 2012 violent scene detection task. In: MediaEval 2012, multimedia benchmark workshop
36. Rumelhart DE, Hinton GE, Williams RJ (1986) Learning representations by back-propagating errors. Nature 323:533–536
37. Safadi B, Quéenot G (2011) Lig at mediaeval 2011 affect task: use of a generic method. In: MediaEval 2011, multimedia benchmark, workshop
38. Schlüter J, Ionescu B, Mironică I, Schedl M (2012) Arf @ mediaeval 2012: an uninformed approach to violence detection in hollywood movies. In: MediaEval 2012, multimedia benchmark, workshop
39. Violence (1996) A public health priority. Technical Report, World Health Organization, Geneva, WHO/EHA/SPI.POA.2
40. Zajdel W, Krijnders JD, Andringa T, Gavrila DM (2007) Cassandra: audio-video sensor fusion for aggression detection. In: IEEE conference on advanced video and signal based surveillance (AVSS 2007), pp 200–205

Chapter 9
Fusion Techniques in Biomedical Information Retrieval

Alba García Seco de Herrera and Henning Müller

Abstract For difficult cases clinicians usually use their experience and also the information found in textbooks to determine a diagnosis. Computer tools can help them supply the relevant information now that much medical knowledge is available in digital form. A biomedical search system such as developed in the Khresmoi project (that this chapter partially reuses) has the goal to fulfil information needs of physicians. This chapter concentrates on information needs for medical cases that contain a large variety of data, from free text, structured data to images. Fusion techniques will be compared to combine the various information sources to supply cases similar to an example case given. This can supply physicians with answers to problems similar to the one they are analyzing and can help in diagnosis and treatment planning.

9.1 Introduction

Clinicians generally base their decisions for diagnosis and treatment planning on a mixture of acquired textbook knowledge and experience acquired through real-life clinical cases [39]. Therefore, in the medical field, two knowledge types are generally available [32]:

- *explicit knowledge*: to the already well-established and formalized domain knowledge, e.g., textbooks or clinical guidelines;
- *implicit knowledge*: individual expertise, organizational practices, and past cases.

A. García Seco de Herrera (✉) · H. Müller
University of Applied Sciences Western Switzerland (HES-SO), TechnoArk 3,
3960 Sierre, Switzerland
e-mail: alba.garcia@hevs.ch

H. Müller
e-mail: henning.mueller@hevs.ch

When working on a new case that includes images, clinicians analyze a series of images together with contextual information, such as the patient age, gender, and medical history as these data can have an impact on the visual appearance of the images. Since related problems may have similar solutions, clinicians use past situations similar to the current one to determine the diagnosis and potential treatment options, information that is also transmitted in teaching, where typical or interesting cases are discussed and used for research [32, 52]. Thus, the goal of a clinician is often to solve a new problem by making use of previous similar situations and by reusing information and knowledge [1], also called case-based reasoning. The problem can be defined in four steps, known as the four 'res' [16, 32]:

1. Retrieve the most similar case(s) from the collection;
2. Reuse them, and more precisely their solutions, to solve the problem;
3. Revise the proposed solution;
4. Retain the current case in the collection for further use.

In this chapter, we focus on the retrieval step because the retrieval of similar cases from a database can help clinicians to find the needed information [39, 45]. In the retrieval step a search over the documents in the database is performed using the formulation of the information need that can include text and images or image regions. Relevant documents are ranked depending on the degree of similarity to a given query case or the similarity to the information need. The most relevant cases are then proposed on the top of the list and can be used to solve the current problem [4].

Text analysis and retrieval has been successfully used in various medical fields from lung disease, through cardiology, eating disorders, to diabetes and Alzheimer's disease [25]. Text in the anamnesis are often the first data available and based on the initial analysis other exams are ordered.

In addition to the text in the anamnesis, another initial data source for diagnosis are the images [52]. Visual retrieval has become an important research area over the past more than 15 years also for medical applications [45]. In the past, the most common visual descriptors used for visual retrieval systems were the color histograms, texture features such as Gabor filters and simple shape measures [45]. In recent years, visual words have had most often the best results in object recognition or image retrieval benchmarks [18] and have become the main way of describing images with a variety of basic features such as SIFT (Scale Invariant Feature Transform) [28] and also texture or color measures.

In terms of medical cases, images are always associated with either text or structured data and this can then be used in addition to the visual content analysis for retrieval. Most often text retrieval has much better performance than visual retrieval, describing the context in which the images were taken. Furthermore, there is an evidence that the combination or fusion of information from textual and visual sources can improve the overall retrieval quality [17, 27]. Whereas visual retrieval usually has good early precision and low recall, text retrieval generally has a high recall.

Combination of image and text search can be done as follows [15]:

- Combine results (ranked lists) of visual and text retrieval for the final results;
- Use visual retrieval to rerank results lists of text retrieval;
- Use text retrieval to rerank results lists of visual retrieval;
- Use image analysis and classification to extract relevant information from the images (such as modality types, anatomic regions, or the recognition of specific objects in the images such as arrows) to filter results lists or rerank them.

In 2013, the Center of Informatics and Information Technology group CITI presented the Nova MedSearch[1] as a medical multimodal (text and image) search engine that can retrieve either similar images or related medical cases [33]. Case-based retrieval taking into account several images and potentially other data of the case has also been proposed by other authors over the past 7 years [37, 52]. Due to the many challenges in biomedical retrieval, research has been attracting increasing attention, and many approaches have been proposed [27].

The remainder of the chapter is organized as follows. Section 9.2 describes the text and visual retrieval and discusses several fusion approaches. The biomedical task and a evaluation framework are presented in Sect. 9.3. In Sect. 9.4 the Khresmoi system is presented as well as the experiments carried out on existing fusion techniques to combine multiple sources. Finally, conclusions are given in Sect. 9.5.

9.2 Visual and Text Information Retrieval

To search through the large amount of data available there is a need for tools and techniques that effectively filter and automatically extract information from text and visual information. Text-based and visual-based methods [22] can in our scenario be used for the retrieval.

9.2.1 Text Retrieval

Most biomedical search engines, also systems searching for images, have been based on text retrieval, only. Sources of biomedical information can be scientific articles and also reports from the patient record [47]. The various parts of the text such as title, abstract, figure captions can then be indexed separately. Some examples for general search tools that have also been used in the biomedical domain are the Lucene, Essie, or Terrier information retrieval (IR) libraries. Lucene[2] is a open source full-text search engine. The advantage of Lucene is its simplicity and high performance [31]. Lucene was chosen for the experiments shown in Sect. 9.4 because it is fast and

---

[1] http://medical.novasearch.org/
[2] http://lucene.apache.org/

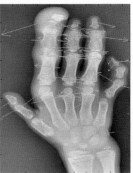

Fig. 9.1 Shape and SIFT (Scale Invariant Feature Transform) information can be extracted from the visual content of the images. In the *left*, the regions detected by a key-region detector are shown. In the *right*, the *arrows* represent the center, scale, and orientation of the keypoints detected by the SIFT algorithm

easy to install and use. Essie [23] is a phrase-based search engine with term and concept query expansion and probabilistic relevancy ranking. It was also designed to use terms from the Unified Medical Language System (UMLS). Terrier[3] is also an open source platform for research and experimentation in text retrieval developed at the University of Glasgow. It supports most state-of-the-art retrieval models such as Dirichlet prior language models, divergence from randomness (DFR) models, or Okapi BM25.

9.2.2 Visual Retrieval

Users of biomedical sources are also often interested in images for biomedical research or medical practice [38], as the images carry an important part of the information in articles. Rather than using text queries, in content-based image retrieval systems, images are indexed and retrieved based on their visual content (image features) such as color, texture, shape, and spatial location of image elements. This allows to use visual information to find images in a database similar to examples given or with similar regions of interest. Figure 9.1 shows examples of the visual information that can be extracted from the images.

The most commonly used features for visual retrieval can be grouped into the following types [22]:

- *Color*: Several color image descriptors have been proposed [5] such as simple color histograms, a color extension to the Scale Invariant Feature Transform (SIFT) [42] or the Bag of Colors [19];
- *Texture*: Texture features have been used to study the spacial organization of pixel values of an image like first-order statistics, second-order statistics, higher order statistics, and multiresolution techniques such as wavelet transform [43].
- *Shape*: Various features have been used to describe shape information, including moments, curvature, or spectral features [53].

---

[3] http://terrier.org/

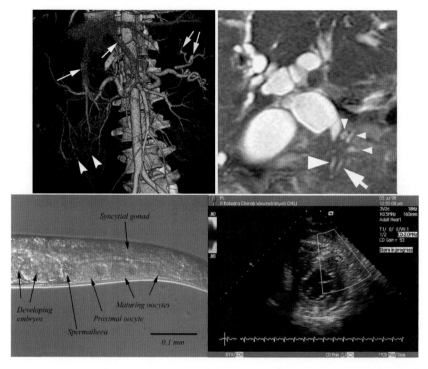

Fig. 9.2 Examples of images with annotations found in the ImageCLEFmed database. Annotations emphasize specific regions of the image according to special attributes of the *highlight region* such as lesions or structures important for the case

Some systems, such as img (Anaktisi),[4] FIRE[5] (Flexible Image Retrieval Engine) or LIRE[6] (Lucene Image REtrieval), allow content-based image retrieval by various visual descriptors and various combinations of descriptors.

In the following, we present some of the processing steps that can potentially improve the retrieval quality of images from the biomedical literature when fusing them with text and/or visual retrieval, particularly for retrieval from the biomedical literature: region-of-interest (ROI) identification, image classification, and multipanel figure separation methods.

9.2.2.1 Region-of-Interest Identification

Annotations in images such as arrows are frequently used in images in the biomedical literature (see Fig. 9.2). If the marked regions can then be linked with text describing

---

[4] http://orpheus.ee.duth.gr/anaktisi/

[5] http://thomas.deselaers.de/fire/

[6] http://www.lire-project.net/

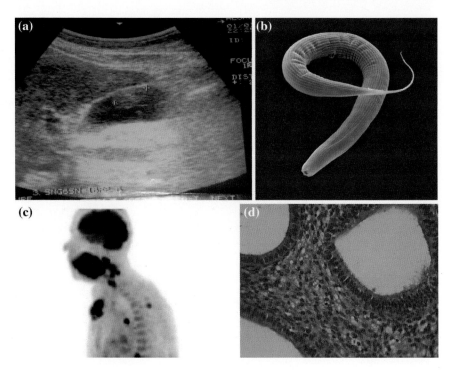

Fig. 9.3 Examples of images of various types that can be found in the biomedical literature **a** Ultrasound **b** Electron microscopy **c** PET **d** Light microscopy

the images, this can be used for retrieval of focused parts of image [40], so retrieving the regions of interest and not entire images. Several approaches have been used in the literature. For instance, Cheng et al. [10] segmented arrow candidates by a global thresholding-based method followed by edge detection. Also Seo et al. [44] developed a semantic ROI segmentation. An attention window is created and a quad-tree-based ROI segmentation is also applied to remove meaningless regions.

9.2.2.2 Image Categorization

In the biomedical literature images can be of several types, some of which correspond to medical imaging modalities such as ultrasound, magnetic resonance imaging (MRI), X-ray, and computer tomography (CT) (see examples in Fig. 9.3). Detecting the image type automatically can help in the retrieval process to focus, for example, on one modality or to remove nonclinical images entirely from the retrieval. Image categories can be integrated into any retrieval system to enhance or filter its results [49], improving the precision of the search [24] and reducing the search space to a set of relevant categories [41]. Furthermore, classification methods can be used to offer adaptive search methods [51]. Using image types as a filter is often requested by clinicians as an important functionality of a retrieval system [30]. Some

9 Fusion Techniques in Biomedical Information Retrieval

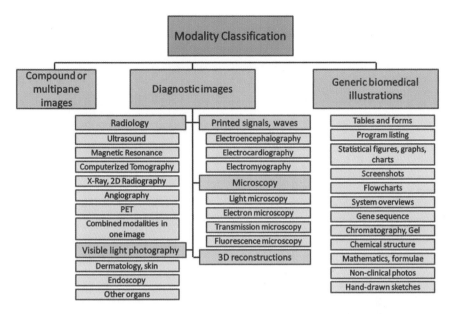

Fig. 9.4 The image class hierarchy proposed in ImageCLEFmed 2013 campaign for the image classification task

Web-accessible retrieval systems such as Goldminer[7] or Yottalook[8] allow users to filter the search results by modality [36].

ImageCLEF[9] proposes a hierarchy of image types for document images occurring in the biomedical open access literature [17], Fig. 9.4 shows the proposed hierarchy. For more details on the ImageCLEF campaign see Sect. 9.3.2. Once the image type information is extracted, the predicted types can be integrated into the search results to generate a final result list. Information on image types can be used in various ways in the retrieval. The following approaches have been used to integrate the classification into the results [49]:

- *Filtering*: Discarding the images of which the predicted type is different to the query. Thus, when filtering using the image type only potentially relevant results are considered;
- *Reranking*: Reranking the initial results with the image type information. The goal is to improve the retrieval ranking by moving relevant documents toward the top of the list based on the categorization;
- *Score fusion*: Fusing a preliminary retrieval score S_R with an image classification score S_M using a weighted sum: $\alpha \cdot S_T + (1 - \alpha) \cdot S_M$, where S_R and S_T are normalized. This approach allows to adjust the parameter α to emphasize the retrieval score or the categorization results.

---

[7] http://goldminer.arrs.org/
[8] http://www.yottalook.com/
[9] http://imageclef.org/

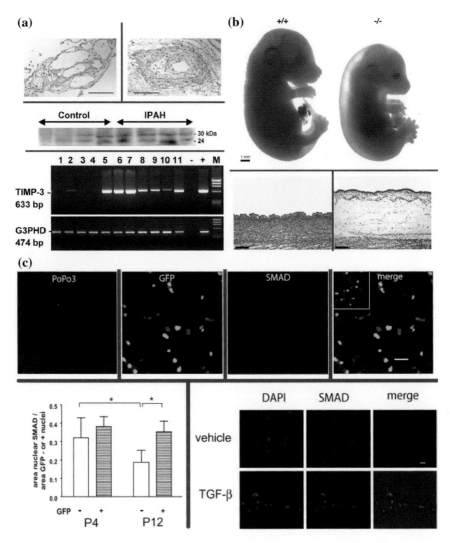

Fig. 9.5 Examples of compound figures found in the ImageCLEFmed database. These examples show mixed modalities in a single figure and several images from the same modality in the same figure. *Red lines* separate the subfigures. **a** Microscopy and chromatography; **b** Mixed modalities; **c** Graphs and microscopy

9.2.2.3 Compound Figure Separation

Compound or multipanel figures (figures consisting of several sub figures) constitute a very large proportion of the images found in the biomedical literature. Image retrieval systems should be capable of distinguishing the parts of compound figures that are relevant to a given query. Compound figure separation is, therefore, a required first step to retrieving focused figures [17]. Figure 9.5 contains several examples of compound figures.

Several approaches have been published for separating figures from text in scanned documents [12] and specifically, for separating compound figures in the biomedical literature [2, 9, 11]. Chhatkuli et al. [11] proposed a compound figure separation technique based on systematic detection and analysis of uniform space gaps. Demner-Fushman et al. [13] determined if an image contains homogeneous regions that cross the entire image. An hybrid clustering algorithm based on particle swarm optimization with a fuzzy logic controller was presented by Cheng et al. [9] to locate related figure components. Using a figure and its associated caption, Apostolova et al. [2] determined if the figure consisted of multiple panels to then separate the panels and the corresponding caption part.

9.2.3 Fusion of Multimodal Features

To combine visual and text search several fusion techniques can be used. Such combinations can lead to better results than single modalities. Text retrieval often has much better performance than visual retrieval in medical retrieval [17], therefore the right combination strategies need to be chosen to really improve performance. This section describes several approaches for information fusion that have been used in the past [14]. To combine the results/features of multiple query images into a single ranked list two main fusion strategies were used depending on how the multiple results from the feature extraction are integrated: early and late fusion. Early fusion integrates unimodal features before making any decision (see Fig. 9.6). Since the decision is then based on all information sources, it enables a truly multimodal feature representation [48]. Unimodal feature vectors are concatenated into one vector using a weighting scheme. Rocchio's algorithm can also be applied to merge the vectors of the same feature spaces into a single vector.

$$\mathbf{q}_m = \alpha \mathbf{q}_o + \beta \frac{1}{|I_r|} \sum_{\mathbf{i}_j \in I_r} \mathbf{i}_j - \gamma \frac{1}{|I_{nr}|} \sum_{\mathbf{i}_j \in I_{nr}} \mathbf{i}_j$$

where α, β and γ are weights, $\mathbf{i}_m$ is the modified query, $\mathbf{i}_o$ is the original query, I_r is the set of relevant documents/images and I_{nr} is the set of nonrelevant documents/images. Only the second term of the right part of the equation is used to merge vectors when nonrelevant documents/images are concerned [18].

Late fusion consists of a combination of independent results from various approaches, e.g., text and visual approaches. The ranked lists of retrieval results are fused and not the features (see Fig. 9.7).

Two main categories of late fusion techniques exist based on which information is used, namely score-based and rank-based methods. In order to obtain a final ranking of a document d fusion techniques are required to reorder documents based on various

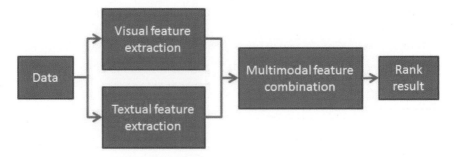

Fig. 9.6 General scheme for early fusion

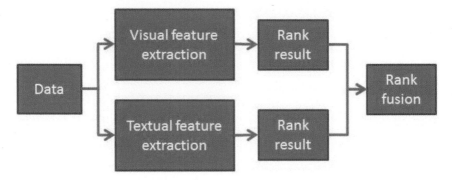

Fig. 9.7 General scheme for late fusion

descriptor lists. An overview of fusion techniques commonly used for the biomedical domain is given below:

- *Score-based methods*:
  - Linear combination
  $$\text{LN}(d) = \alpha S_t(d) + \beta S_v(d)$$
  where S_t and S_v are the textual and visual scores of the document d;
  - combSUM
  $$\text{combSUM}(d) = \sum_{j=1}^{N_j} S_j(d)$$
  with N_j being the number of descriptors to be combined and $S(i)$ is the score assigned to document d;
  - combMNZ
  $$\text{combMNZ}(d) = F(d) * \text{combSUM}(d)$$

where $F(d)$ is the frequency of document d being returned by one input system with a nonzero score;
- combMAX
$$\text{combMAX}(d) = \arg\max_{j=1:N_j}(S_j(d))$$

- combMIN
$$\text{combMIN}(d) = \arg\min_{j=1:N_j}(S_j(d))$$

- combPROD
$$\text{combPROD}(d) = \prod_{j=1}^{N_j} S_j(d)$$

- *Rank-based methods*:
  - Reciprocal rank fusion:
  $$\text{RRFscore}(d) = \sum_{r \in R} \frac{1}{k + r(d)}$$

  where R is the set of rankings assigned to the documents;
  - Borda
  $$\text{Borda}(d) = \sum_{r \in R} r(d).$$

9.3 Biomedical Retrieval

In this section, a biomedical retrieval scenario is investigated. An evaluation framework for biomedical retrieval systems is proposed by ImageCLEFmed and this chapter uses the same framework to make results comparable with the state of the art.

9.3.1 Medical Scenario

A biomedical retrieval system should correspond to real practical informations and be evaluated based on a corresponding scenario. Several user surveys and analyses of search log files have been done for obtaining the place of text and visual information in retrieval, mainly in radiology [34, 35, 50]. Based on these user analyses the tasks in ImageCLEFmed were developed. Particularly, radiologists frequently search for images and have a need to search for visual abnormalities linked to specific pathologies. Usually, not the entire image is of interest but rather small regions of interest [46]. In the past, text retrieval for these tasks has obtained much better information that visual retrieval [17], but combinations can profit from the advantages of the two.

Currently, the use of only visual information still achieves low retrieval performance in this task and the combination of text and visual search is improving and seems promising [17].

9.3.2 ImageCLEFmed: An Evaluation Framework

ImageCLEF[10] is the image retrieval track of the Cross Language Evaluation Forum (CLEF)[11] [6]. One of the main goals of the medical task of ImageCLEF (ImageCLEFmed) [17] is to investigate the effectiveness of combining text and images for medical image- and case-based retrieval [14]. Several tasks have been proposed over the years since 2004, always in a very end user-oriented way based on surveys or log files analyses. In 2013, four tasks were organized:

- Image-based retrieval;
- Case-based retrieval;
- Modality classification;
- Compound figure separation.

The image-based retrieval task has been running since 2004 with changing databases. The goal of this task is to retrieve images for a precise information need expressed through text and example images. Figure 9.8 shows one of the 35 topics distributed to the participants in 2013.

The case-based task was first introduced in 2009. This task aims to retrieve cases that are similar to the query case and are useful in differential diagnosis. Each topic consists of a case description with patient demographics, limited symptoms and test results including imaging studies (but not the final diagnosis). An example of a topic can be seen in Fig. 9.9.

Since 2010, the modality classification task has been running. The goal of this task is to classify images into image types that can be medical modalities or other types occurring in the biomedical literature. More information can be found in Sect. 9.2.2.2.

In 2013, a compound figure separation task was added as a large portion of images in the literature turn out to be compound figures.

The ImageCLEFmed evaluation framework gives access to the tools developed for the described tasks including databases and ground truth. These tools were used to conduct the fusion experiments presented in Sect. 9.4.

---

[10] http://imageclef.org/
[11] http://www.clef-initiative.eu/

9 Fusion Techniques in Biomedical Information Retrieval

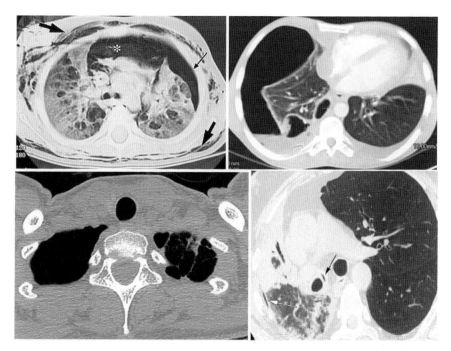

Fig. 9.8 Images from one of the topics in the image-based retrieval task of ImageCLEFmed 2013. They correspond to the textual query "pneumothorax CT images" that is also expressed in French, German, and Spanish

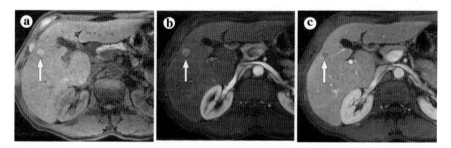

Fig. 9.9 Images from one of the topics in the case-based retrieval task of ImageCLEFmed 2013. They correspond to the textual query "A 56-year-old woman with Hepatitis C, now with abdominal pain and jaundice. Abdominal MRI shows T1 and T2 hyperintense mass in the left lobe of the liver which is enhanced in the arterial phase"

Table 9.1 Results of the approaches at the ImageCLEF case-based retrieval task when using only text retrieval

Run	MAP	Bpref	P10	P30
Best textual ImageCLEF run	0.2429	0.2417	0.2657	0.1981
Textual	0.1791	0.1630	0.2143	0.1581

9.4 Khresmoi and Evaluation of Fusion Techniques

To search through large amounts of biomedical data, the Khresmoi[12] project is developing a multilingual multimodal search and access system for medical and health information and documents [3].

In this section, the work on text and visual fusion as part of Khresmoi is presented. More on the employed fusion techniques can also be found in [21]. The experiments use the ImageCLEFmed 2013 database of the case-based task described in Sect. 9.3.2.

For text retrieval, Apache Lucene was used (see Sect. 9.2.1). The results achieved with this approach on the case-based task of ImageCLEF is shown in Table 9.1.

For visual retrieval, a combination of the following descriptors were extracted to incorporate color and texture information from the images [20]:

- color and edge directivity descriptor (CEDD) [7];
- bag of visual words using SIFT, Scale Invariant Feature Transform, (BoVW) [28];
- fuzzy color and texture histogram (FCTH) [8];
- bag of colors (BoC) [19];
- BoVW with a spatial pyramid matching [26] (BoVW-SPM);
- BoC with $n \times n$ spatial grid (Grid BoC).

To enhance visual retrieval several fusion strategies described in Sect. 9.2.3 were tested to combine results of each of the query images and of several visual descriptors of the same image. Table 9.2 shows the results of the visual retrieval using this combination of fusion rules.

The results of the combination of text and visual approach are shown in Table 9.3. The visual approach selected for these combination used RRF for query fusion and combSUM for the fusion of the descriptors, obtaining the best results in terms of MAP (MAP $= 0.0037$) and $P30$ ($P30 = 0.0067$) (see Table 9.2).

Although in previous ImageCLEF campaigns, the mixed submissions sometimes achieved worse performance than the textual runs, the best result among all the experiments carried out on this chapter was obtained using a linear combination of text and visual search (MAP $= 0.1795$). The weight of each rank was defined by a function of their performance in terms of MAP, where the best MAP scores obtained using text (MAP$(T) = 0.1293$) and visual (MAP$(V) = 0.0204$) search in ImageCLEFmed 2011 [29] were employed. Despite the lower performance of the visual and textual approaches compared with the runs submitted to ImageCLEFmed 2013, the fusion of visual and text retrieval outperform the best multimodal approach

---

[12] http://www.khresmoi.eu/

Table 9.2 Results of the approaches for the ImageCLEF case-based retrieval task when using various fusion strategies for visual retrieval. Query and descriptor fusion is combined

Queries fusion	Descriptors fusion	MAP	Bpref	P10	P30
Best visual	ImageCLEF run	0.0281	0.0335	0.0429	0.0238
Rocchio	Borda	0.0004	0.0092	0	0
Rocchio	combMAX	0.0004	0.0096	0	0.0029
Rocchio	combMIN	0.0002	0.0093	0	0.0019
Rocchio	combMNZ	0.0008	0.0084	0.0029	0.0048
Rocchio	combSUM	0.0006	0.0084	0.0029	0.0038
Rocchio	RRF	0.0005	0.0085	0	0.0038
Borda	Borda	0.0005	0.0060	0	0.0019
Borda	combMAX	0.0004	0.0066	0	0.0019
Borda	combMIN	0.0002	0.0124	0	0
Borda	combMNZ	0.0009	0.0055	0.0029	0.0038
Borda	combSUM	0.0005	0.0060	0.0029	0.0029
Borda	RRF	0.0012	0.0061	0.0086	0.0057
combMAX	Borda	0.0006	0.0062	0.0066	0.0019
combMAX	combMAX	0.0006	0.0089	0.0057	0.0029
combMAX	combMIN	0.0003	**0.0156**	0	0.0019
combMAX	combMNZ	0.0036	0.0077	**0.0114**	0.0057
combMAX	combSUM	0.0021	0.0077	0.0086	**0.0067**
combMAX	RRF	0.0013	0.0066	0.0086	0.0048
combMIN	Borda	0.0005	0.0077	0.0029	0.0029
combMIN	combMAX	0.0006	0.0091	0.0086	0.0038
combMIN	combMIN	0.0003	0.0172	0	0.0019
combMIN	combMNZ	0.0032	0.008	0.0086	0.0057
combMIN	combSUM	0.0015	0.0079	0.0057	0.0057
combMIN	RRF	0.0011	0.0060	0.0086	**0.0067**
combMNZ	Borda	0.0005	0.0061	0.0029	0.001
combMNZ	combMAX	0.0004	0.0077	0	0.0038
combMNZ	combMIN	0.0001	0.0111	0	0.001
combMNZ	combMNZ	0.0029	0.0058	0.0086	**0.0067**
combMNZ	combSUM	0.0011	0.0053	0.0057	0.0057
combMNZ	RRF	0.0008	0.0055	0.0029	0.0038
combSUM	Borda	0.0005	0.006	0.0029	0.0019
combSUM	combMAX	0.0005	0.0084	0.0057	0.0038
combSUM	combMIN	0.0002	0.0127	0	0.0019
combSUM	combMNZ	0.0033	0.0075	0.0086	0.0076
combSUM	combSUM	0.0014	0.0067	0.0086	**0.0067**
combSUM	RRF	0.0009	0.0051	0.0029	0.0048
RRF	Borda	0.0005	0.0057	0	0.0019
RRF	combMAX	0.0004	0.0070	0	0.0038
RRF	combMIN	0.0002	0.0121	0	0
RRF	combMNZ	**0.0037**	0.0129	0.0086	**0.0067**
RRF	combSUM	0.0011	0.0060	0.0086	**0.0067**
RRF	RRF	0.0010	0.0047	0.0029	0.0057

Table 9.3 Results of the approaches for the ImageCLEF case-based retrieval task when using various fusion strategies to combine visual and textual information

Visual + textual fusion	MAP	Bpref	P10	P30
Best ImageCLEF run	0.1608	0.1426	0.1800	0.1257
Borda	0.1302	0.1230	0.1371	0.1105
combMAX	0.1770	0.1625	0.2143	**0.1571**
combMIN	0.1505	0.157	0.2171	0.1438
combMNZ	0.1197	0.1257	0.1714	0.1133
combSUM	0.1741	0.1609	**0.2229**	0.161
RRF	0.1084	0.1011	0.1543	0.1114
LN	**0.1795**	**0.1627**	0.2086	**0.1571**

submitted to ImageCLEFmed 2013. This shows the importance of the multimodal fusion.

Sections 9.2.2.2 and 9.2.2.3 describe the modality classification and compound figure separation tasks. Both can be fused with text and visual retrieval to improve the quality of the systems but this has not yet been implemented in our approach. The modality classification of Khresmoi approached achieved an accuracy of 69.63 %. Moreover, the compound figure separation approach obtained the best accuracy of all the ImageCLEF 2013 participants (84.64 %) [20]. There seems to be potential for improving performance including these techniques into the retrieval process.

9.5 Conclusions

In their practical work, clinicians have information needs when taking informed decisions. Their work sometimes involves search for similar past cases. To help clinicians in their daily routine, several information retrieval approaches involving visual and text retrieval are proposed.

This chapter describes the methods to combine both visual and textual information in a biomedical retrieval system. In the context of the Khresmoi project, experiments on text and visual fusion were done using the ImageCLEFmed 2013 database. Despite the low performance of the visual approaches on the case-based task, the fusion of text and visual techniques are improving the quality of the retrieval. Applying weighted linear combination of text and visual retrieval ranks, results outperform the best multimodal runs submitted at ImageCLEFmed 2013 with a MAP of 0.1795. It demonstrates the effectiveness of proposed the multimodal framework. Moreover, image analysis can be applied to enhance the quality of retrieval system. Image classification and compound figure separation are common techniques that can be integrated into a retrieval systems to improve the performance of a simple text or visual retrieval. The system achieved and accuracy of 69.63 % at the ImageCLEFmed 2013 modality classification task and 84.64 % at the compound figure separation task.

Future work will focus on integration the modality classification and compound figure separation into the retrieval system to show that they can contribute to improve the retrieval.

Acknowledgments This work was partly supported by the EU 7th Framework Program in the context of the Khresmoi project (FP7-257528).

References

1. Aamodt A, Plaza E (1994) Case-based reasoning: foundational issues, methodological variations, and systems approaches. AIC 7(1):39–59
2. Apostolova E, You D, Xue Z, Antani S, Demner-Fushman D, Thoma GR (2013) Image retrieval from scientific publications: text and image content processing to separate multi-panel figures. J Am Soc Inf Sci Technol 64(5):893–908
3. Aswani N, Beckers T, Birngruber E, Boyer C, Burner A, Bystron J, Choukri K, Cruchet S, Cunningham H, Dedek J, Dolamic L, Donner R, Dungs S, Eggel I, Foncubierta-Rodríguez A, Fuhr N, Funk A, García Seco de Herrera A, Gaudinat A, Georgiev G, Gobeill J, Goeuriot L, Gómez P, Greenwood M, Gschwandtner M, Hanbury A, Hajic J, Hlaváčová J, Holzer M, Jones G, Jordan B, Jordan M, Kaderk K, Kainberger F, Kelly L, Mriewel S, Kritz M, Langs G, Lawson N, Markonis D, Martinez I, Momtchev V, Masselot A, Mazo H, Müller H, Pecina P, Pentchev K, Peychev D, Pletneva N, Pottecherc D, Roberts A, Ruch P, Samwald M, Schneller P, Stefanov V, Tinte MA, Uresová Z, Vargas A, Vishnyakova D (2012) Khresmoi: multimodal multilingual medical information search. In: Proceedings of the 24th international conference of the European federation for medical informatics
4. Begum S, Ahmed MU, Funk P, Xiong N, Folke M (2011) Case-based reasoning systems in the health sciences: a survey of recent trends and developments. IEEE Trans Syst Man Cybern 41(4):421–434
5. Burghouts GJ, Geusebroek JM (2009) Performance evaluation of local colour invariants. Comput Vis Image Underst 113(1):48–62
6. Caputo B, Müller H, Thomee B, Villegas M, Paredes R, Zellhofer D, Goeau H, Joly A, Bonnet P, Martinez Gomez J, Garcia Varea I, Cazorla C (2013) ImageCLEF 2013: the vision, the data and the open challenges. In: Working notes of CLEF 2013 (Cross Language Evaluation Forum)
7. Chatzichristofis SA, Boutalis YS (2008) Cedd: color and edge directivity descriptor: a compact descriptor for image indexing and retrieval. Lect Notes Comput Sci 5008:312–322
8. Chatzichristofis, SA, Boutalis YS (2008) FCTH: fuzzy color and texture histogram: a low level feature for accurate image retrieval. In: Proceedings of the 9th international workshop on image analysis for multimedia interactive service, pp 191–196
9. Cheng B, Sameer A, Stanley RJ, Thoma GR (2011) Automatic segmentation of subfigure image panels for multimodal biomedical document retrieval. In: Agam G, Viard-Gaudin C (eds.) Document recognition and retrieval. SPIE Proceedings, SPIE, vol 7874, pp 1–10
10. Cheng B, Stanley RJ, De S, Antani S, Thoma GR (2011) Automatic detection of arrow annotation overlays in biomedical images. Int J Healthc Inf Syst Inform 6(4):23–41
11. Chhatkuli A, Markonis D, Foncubierta-Rodríguez A, Meriaudeau F, Müller H (2013) Separating compound figures in journal articles to allow for subfigure classification. In: SPIE medical imaging
12. Chiu P, Chen F, Denoue L (2010) Picture detection in document page images. In: Proceedings of the 10th ACM symposium on document engineering, pp 211–214, ACM
13. Demner-Fushman D, Antani S, Simpson MS, Thoma GR (2012) Design and development of a multimodal biomedical information retrieval system. J Comput Sci Eng 6(2):168–177

14. Depeursinge A, Müller H (2010) Fusion techniques for combining textual and visual information retrieval. In: Müller H, Clough P, Deselaers T, Caputo B (eds) ImageCLEF, The Springer international series on information retrieval, vol 32. Springer, Berlin Heidelberg, pp 95–114
15. Glasgow J, Jurisica I (1998) Integration of case-based and image-based reasoning. In: AAAI workshop on case-based reasoning integrations. AAAI Press, Menlo Park, California, pp 67–74
16. Gu M, Aamodt A, Tong X (2005) Component retrieval using conversational case-based reasoning. In: Intelligent information processing II. Springer-Verlag, London, UK, pp 259–271
17. García Seco de Herrera A, Kalpathy-Cramer J, Demner Fushman D, Antani S, Müller H (2013) Overview of the ImageCLEF 2013 medical tasks. In: Working notes of CLEF 2013 (Cross Language Evaluation Forum)
18. García Seco de Herrera A, Markonis D, Eggel I, Müller H (2012) The medGIFT group in ImageCLEFmed 2012. In: Working notes of CLEF 2012
19. García Seco de Herrera A, Markonis D, Müller H (2013) Bag of colors for biomedical document image classification. In: Greenspan H, Müller H (eds) Medical content-based retrieval for clinical decision support. In: MCBR-CDS 2012. Lecture Notes in Computer Sciences (LNCS), pp 110–121
20. García Seco de Herrera A, Markonis D, Schaer R, Eggel I, Müller H (2013) The medGIFT group in ImageCLEFmed 2013. In: Working notes of CLEF 2013 (Cross Language Evaluation Forum)
21. García Seco de Herrera A, Schaer R, Müller H (submitted) Comparing fusion techniques for the ImageCLEF 2013 medical case retrieval task. Comput Med Imaging Graph
22. Hwang HK, Lee H, Choi D (2012) Medical image retrieval: past and present. Healthc Inf Res 18(1):3–9
23. Ide NC, Loane RF, Demner-Fushman D (2007) Essie: a concept-based search engine for structured biomedical text. J Am Med Inform Assoc 14(3):253–263
24. Kalpathy-Cramer J, Hersh W (2010) Multimodal medical image retrieval: image categorization to improve search precision. In: Proceedings of the international conference on multimedia information retrieval, MIR '10ACM, New York, NY, USA, pp 165–174
25. Kwiatkowska M, Atkins S (2004) Case representation and retrieval in the diagnosis and treatment of obstructive sleep apnea: a semiofuzzy approach. In: Proceedings European case based reasoning conference, ECCBR'04
26. Lazebnik S, Schmid C, Ponce J (2006) Beyond bags of features: spatial pyramid matching for recognizing natural scene categories. In: Proceedings of the 2006 IEEE conference on computer vision and pattern recognition, CVPRIEEE Computer Society, Washington, DC, USA, pp 2169–2178
27. Li Y, Shi N, Frank DH (2011) Fusion analysis of information retrieval models on biomedical collections. In: Proceedings of the 14th international conference on information fusion. IEEE Computer Society
28. Lowe DG (2004) Distinctive image features from scale-invariant keypoints. Int J Comput Vis 60(2):91–110
29. Markonis D, García Seco de Herrera A, Eggel I, Müller H (2011) The medGIFT group in ImageCLEFmed 2011. In: Working notes of CLEF 2011
30. Markonis D, Holzer M, Dung S, Vargas A, Langs G, Kriewel S, Müller H (2012) A survey on visual information search behavior and requirements of radiologists. Methods Inf Med 51(6):539–548
31. McCandless M, Hatcher E, Gospodnetic O (2010) Lucene in action, second edition: Covers Apache Lucene 3.0. Manning Publications, Greenwich, CT, USA
32. Montani S, Bellazzi R (2002) Supporting decisions in medical applications: the knowledge management perspective. Int J Med Inform 68:79–90
33. Mourão A, Martins F (2013) NovaMedsearch: a multimodal search engine for medical case-based retrieval. In: Proceedings of the 10th conference on open research areas in information retrieval, OAIR'13, pp 223–224
34. Müller H, Boyer C, Gaudinat A, Hersh W, Geissbuhler A (2007) Analyzing web log files of the health on the Net HONmedia search engine to define typical image search tasks for image

retrieval evaluation. MedInfo 2007, vol 12. Studies in health technology and informatics. IOS press, Brisbane, Australia, pp 1319–1323
35. Müller H, Despont-Gros C, Hersh W, Jensen J, Lovis C, Geissbuhler A (2006) Health care professionals' image use and search behaviour. Proceedings of the medical informatics Europe conference (MIE 2006). Studies in health technology and informatics. IOS Press, Maastricht, The Netherlands, pp 24–32
36. Müller H, García Seco de Herrera A, Kalpathy-Cramer J, Demner Fushman D, Antani S, Eggel I (2012) Overview of the ImageCLEF 2012 medical image retrieval and classification tasks. In: Working notes of CLEF 2012 (Cross Language Evaluation Forum)
37. Müller H, Zhou X, Depeursinge A, Pitkanen M, Iavindrasana J, Geissbuhler A (2007) Medical visual information retrieval: state of the art and challenges ahead. In: Proceedings of the 2007 IEEE international conference on multimedia and Expo, ICME'07, IEEE, pp 683–686
38. Philip A, Afolabi B, Oluwaranti A, Oluwatolani O (2011) Development of an image retrieval model for biomedical image databases. In: Jao C (ed) ISBN: 978-953-307-258-6, InTech, Available from: http://www.intechopen.com/books/efficient-decision-support-systems-practice-and-challenges-in-biomedical-related-domain/development-of-an-image-retrieval-model-for-biomedical-image-databases
39. Quellec G, Lamard M, Bekri L, Cazuguel G, Roux C, Cochener B (2010) Medical case retrieval from a committee of decision trees. IEEE Trans Inf Technol Biomed 14(5):1227–1235
40. Rahman MM, You D, Simpson MS, Antani S, Demner-Fushman D, Thoma GR (2012) An interactive image retrieval framework for biomedical articles based on visual region-of-interest (ROI) identification and classification. In: Proceedings of the IEEE second international conference on healthcare informatics, imaging and systems biology, HISB
41. Rahman MM, You D, Simpson MS, Antani SK, Demner-Fushman D, Thoma GR (2013) Multimodal biomedical image retrieval using hierarchical classification and modality fusion. Int J Multimedia Inf Retrieval 2(3):159–173
42. van de Sande KEA, Gevers T, Smeulders AWM (2010) The university of amsterdam's concept detection system at imageclef 2009. Lect Notes Comput Sci 6242:261–268
43. Selvarajah S, Kodituwakku SR (2011) Analysis and comparison of texture features for content based image retrieval. Int J Latest Trends Comput 2:108–113
44. Seo M, Ko B, Chung H, Nam J (2006) ROI-based medical image retrieval using human-perception and MPEG-7 visual descriptors. Proceedings of the 5th international conference on image and video retrieval, CIVR'06. Springer-Verlag, Berlin, Heidelberg, pp 231–240
45. Shapiro LG, Atmosukarto I, Cho H, Lin HJ, Ruiz-Correa S, Yuen J (2008) Similarity-based retrieval for biomedical applications. In: Case-based reasoning on images and signals, Studies in computational intelligence, vol 73. Springer, pp 355–387
46. Simonyan K, Modat M, Ourselin S, Criminisi A, Zisserman A (2013) Immediate ROI search for 3-D medical images. In: Greenspan H, Müller H (eds) Medical content-based retrieval for clinical decision support. In: MCBR-CDS 2012. Lecture Notes in Computer Sciences (LNCS)
47. Simpson MS, Demner-Fushman D (2012) Biomedical text mining: a survey of recent progress. In: Aggarwal CC, Zhai C (eds) Mining text data. Springer, pp 465–517
48. Snoek CGM, Worring M, Smeulders AWM (2005) Early versus late fusion in semantic video analysis. In: MULTIMEDIA '05: Proceedings of the 13th annual ACM international conference on multimedia, pp 399–402. ACM, New York, NY, USA
49. Tirilly P, Lu K, Mu X, Zhao T, Cao Y (2011) On modality classification and its use in text-based image retrieval in medical databases. In: 9th international workshop on content-based multimedia indexing
50. Tsikrika T, Müller H, Kahn Jr, CE (2012) Log analysis to understand medical profession-als' image searching behaviour. In: Proceedings of the 24th European medical informatics conference, MIE'2012
51. Wang JZ (2000) Region-based retrieval of biomedical images. In: Proceedings of the ACM multimedia conference, pp 511–512
52. Welter P, Deserno TM, Fischer B, Günther RW, Spreckelsen C (2011) Towards case-based medical learning in radiological decision making using content-based image retrieval. BMC Med Inform Decis Mak 11:68

53. Zhang D, Lu G (2004) Review of shape representation and description techniques. Pattern Recogn 37(1):1–19

Chapter 10
Using Crowdsourcing to Capture Complexity in Human Interpretations of Multimedia Content

Martha Larson, Mark Melenhorst, María Menéndez and Peng Xu

Abstract Large-scale crowdsourcing platforms are a key tool allowing researchers in the area of multimedia content analysis to gain insight into how users interpret social multimedia. The goal of this article is to support this process in a practical manner that opens the path for productive exploitation of complex human interpretations of multimedia content within multimedia systems. We first discuss in detail the nature of complexity in human interpretations of multimedia, and why we, as researchers, should look outward to the crowd, rather than inward to ourselves, to determine what users consider important about the content of images and videos. Then, we present strategies and insights from our own experience in designing tasks for crowdworkers. Our techniques are useful to researchers interested in eliciting information about the elements and aspects of multimedia that are important in the contexts in which humans use social multimedia.

10.1 Introduction

The world in which we live is complex. It is made up of many interacting elements and aspects. Our understanding of the world is based not only on what is going on, but also on where it is taking place and who is involved. Given the intricacy of this

M. Larson (✉) · M. Melenhorst · P. Xu
Department of Intelligent Systems, Delft University of Technology, Delft, Netherlands
e-mail: m.a.larson@tudelft.nl

M. Melenhorst
e-mail: m.s.melenhorst@tudelft.nl

P. Xu
e-mail: p.xu@tudelft.nl

M. Menéndez
Department of Information Engineering and Computer Science, University of Trento, Trento, Italy
e-mail: menendez@disi.unitn.it

interplay in the world around us, it is unsurprising that when we make multimedia recordings of that world (i.e., take a picture or capture a video) that these are also by nature complex. The purpose of this article is to provide a useful characterization of the nature of this complexity, as well as practical insight that will support researchers in social multimedia in tackling complex human interpretations.

Initially, the decision to address complex human interpretations of multimedia seems to open Pandora's box. Since we do not assume the existence of any principles that would exclude particular human interpretations, we are forced to allow, at least in theory, any interpretation to be assigned to any image or video. Seen in this way, admitting the complexity of human interpretation to the study of multimedia appears to create an unbounded problem.

This contribution of this chpater is to offer a perspective that makes it possible to take the initial steps of confronting complexity without formulating it as an unbounded problem. We advocate that complexity should be first recognized, and then the aspects of complexity that are most important to users should be isolated and tackled. The chapter illustrates some key contributions that can be made by large-scale crowdsourcing platforms, i.e., systems capable of collecting answers to specific questions from a large number of people.

The insights of the article are intended to be useful to allow multimedia researchers to move from focusing their research on simple descriptions of multimedia, to investigating techniques that address descriptions that are more complex and, for this reason, have the potential to be more directly useful to the users of multimedia systems. This chapter, in some respects, can be considered a complete mirror image of much of the literature written in the area of multimedia content analysis. It focuses on what humans do with multimedia, rather than what algorithms can do. It is independent of any particular multimedia content analysis technology: no specific features or automatic analysis algorithms are mentioned. Finally, many of the insights provided by this chapter take the form not of answers, but questions that should be asked when carrying out multimedia research. We choose this form in order to keep the focus of the chapter not on what current technology can accomplish, but rather on productive routes of inquiry. Specifically, we are interested in techniques that have recently become more feasible to pursue, but that we are currently still in danger of overlooking.

It is important to note that in this chapter we focus on human interpretations of the literally depicted content of images and video. Specifically, we are interested in annotations contributed by users that address the questions about an image or a video: "What does this show?" and "What is this about?" In other words, we are interested in "objective" descriptions of multimedia, i.e., descriptions that focus on the object in question, the "what". The chapter includes a comparison with "subjective" descriptions of multimedia, i.e., descriptions that focus on the reaction of subject, (e.g., the emotional response or esthetic preference of the human user) and discusses the implications of the differences.

The chapter is structured into two parts. The first part (Sects. 10.2–10.3) discusses how humans interpret multimedia. In Sect. 10.2, we explain why handling the complexity of human interpretations is important for multimedia systems, and provide additional examples of contrasting multimedia descriptions involving simple concepts and multimedia descriptions involve complex interpretations. Then, in Sect. 10.3, we delve more deeply into the nature of complex interpretations, including the dependency of interpretation on context and the relationship between complexity and other notions discussed in the literature. The second part of the chapter (Sects. 10.4–10.5) discusses crowdsourcing and its role in eliciting complex interpretations of multimedia from large groups of people. In Sect. 10.4, we briefly introduce crowdsourcing for multimedia, and discuss its added value. In Sect. 10.5, we cover an example that illustrates crowdsourcing at work to collect complex interpretations. This example leads to a presentation of specific crowdsourcing techniques and a discussion of their limitations. We finish with a conclusion and outlook in Sect. 10.6.

10.2 The Importance of Human Interpretations

We begin with the observation that upon first considerations it seems natural that the nature of human interpretation should be easy to grasp. After all, we are all human and engage in processes of interpretation constantly in our daily lives. However, as it turns out, just like we need a mirror if we want to look at our own eyes, we need external tools to understand our own complex interpretations. In this chapter, we will argue that crowdsourcing is the source of an external tool well suited for this purpose. However, before we turn to crowdsourcing, we discuss in depth why complex human interpretations are important for multimedia systems and exactly what we mean when we say a human interpretation is complex.

10.2.1 Human Interpretations of Multimedia in Multimedia Systems

Multimedia systems are systems that provide users with access to collections of multimedia content, via search, recommendation, or browsing. Such systems are able to take advantage of descriptions of multimedia that encode human interpretations. These descriptions make possible a close match between the needs of users and content in the collection.

A prime example of a multimedia system that exploits human interpretations is an image search engine that uses tags assigned by uploaders sharing their images online in order to match user queries with relevant items. However, although tags and annotations that are contributed by users who upload or interact with content are valuable, they do not fully characterize human interpretations of multimedia. Tags are known to be sparse, and the information that they provide partial, at best. For these reasons, multimedia content analysis, i.e., the automatic generation of multimedia descriptions, holds great promise to improve the usefulness of multimedia systems for users.

Arguably, the most daunting challenge facing researchers in the area of multimedia content analysis is the sheer diversity of multimedia. The diversity is especially challenging in the case of social multimedia on the Internet, whose characteristics are highly unpredictable. As they work to develop methods to bridge the semantic gap, the difference between machine representations (i.e., pixel patterns) and human understanding, researchers struggle to find techniques that are capable of generalizing in the face of wide variability. Much multimedia content analysis research involves addressing diversity on the "machine side" of the semantic gap. For example, visual concept detection attempts to develop methods that can robustly generalize over different visual appearances of instances of entities belonging to the same conceptual class.

However, diversity in multimedia also exists on the "human side" of the semantic gap. Different people will say different things about the same image, when they are asked what the image depicts. Answers vary depending on where or when people are asked. For forms of multimedia content that go beyond images, the same observation holds: people provide different answers when asked what a video is about.

This chapter, as mentioned above, opens relevant questions and supplies some practical ideas and techniques for researchers who are interested in creating multimedia content analysis algorithms that are sensitive to the complexity of human interpretation of multimedia. The ultimate goal is to contribute to improving algorithms that generate automatic descriptions of multimedia content, and are useful for multimedia systems.

10.2.2 Complex Interpretations and Next Generation Multimedia Analysis

If human interpretations of multimedia are complex, it is not surprising that descriptions involving simple concepts fall short of capturing the range of what can be seen in an image or what is shown in a video. Yet, much current work in the area of multimedia focuses on simple concepts. For example, it aims to process images to detect animals from a typical inventory such as would be depicted in a children's picture dictionary.

The appeal of simple concepts for the multimedia community lies in three factors: the likelihood that their encodings are relatively stable (e.g., visually invariant), the likelihood that they are well represented within multimedia collections, and the idea that they can be combined compositionally into more complex representations. We discuss each of these in turn.

Consider the two images in Figure 10.1. Comparison between the images suggests that the depiction of the simple concept "Golden Retriever" is visually similar across images, making it possible to train a concept detector. The Web yields a large number of "Golden Retriever" images. Should there not be enough "Golden Retriever" data to train a good concept detector, it is possible to back off to the more general concept "dog". Some sacrifice must be made on the visual stability across representations of dogs in different images, but it is safe to assume that tens of thousands of images to train a dog detector can be readily obtained from the Web. Finally, it seems relatively straightforward that in order to identify images fitting the description, "Golden Retriever in the snow," a useful approach would be to decompose the problem, i.e., to detect "Golden Retriever" and then detect "snow" and intersect the two result sets.

These considerations have driven much of the work on multimedia content analysis. In particular, we mention the work of Dong Liu et al. [14], in the area of "Tag ranking". This work uses the image on the right in Fig. 10.1 as an example. This article cites the tags that have been assigned by the uploader to this image and states that the most relevant tag is "dog" (cf. [14, p. 351], Fig. 10.1). The tag "dog" occupies the fifth position in the uploader-assigned tag list, and is preceded by the tags, "alex", "meditating", "love", and "winter". Without doubt, it is productive for a researcher in the area of multimedia content analysis to assume that the tag "dog" is most interesting in this list. With high probability, it is for this tag that one could collect the largest amount of relatively visually stable data. This data could be used to train a model capable of automatically generating the label "dog" as a description for this image.

However, there are a number of reasons why it has recently become both interesting and feasible to move beyond using simple concepts to characterize multimedia. First, there has been a growing realization that the perfect "dog" detector is only a partial solution to building a useful multimedia retrieval system. There are thousands of images on the Internet depicting dogs. Focusing on "Golden Retrievers" does not narrow the problem significantly. A system that returns that large number of undifferentiated "dog" or "Golden Retriever" results is not supporting users in closing in on images that have a tight fit with their information needs. The image on the right in Fig. 10.1 is not even a particularly good dog picture, since it does not portray the dog canonically, but rather, the dog's eyes are closed. In sum, the tag "dog" does not provide a particularly satisfying characterization of this image.

Second, with the advent of crowdsourcing, to be discussed in detail later in this chapter, it is no longer necessary to assume that descriptions for images must be generated by fully automatic approaches. Instead hybrid approaches that combine human intelligence and machine computation can be used to generate image descriptions. Hybrid approaches have already been applied with a great deal of success to the case of simple concepts [29]. However, human judgments are more subtle and richer. For a human, it is easy to see in the image on the right in Fig. 10.1 that the dog's eyes are closed, and that its face is illuminated by a light source. These are outward signs that conventionally indicate that a human depicted in an image is meditating. A human interpreting this image realizes instantly that, if the dog is anthropomorphized, the label "meditating" is a good fit. This example is an illustration of a complex interpretation of an image. Since modern systems are able to consult human intelligence in the form of the crowd, in addition to carrying out computation, it is no longer necessary to consider "meditating," or other complex interpretations, to be a less relevant for images. In other words, the semantic gap need no more be considered to represent an absolute barrier.

Fig. 10.1 In the case of social multimedia, a simple concept, like "dog", does not always adequately describe an image. *Flickr credit* Andrea Arden (*Left*) and Andrew Morrell Photography (*Right*)

Third, a priori, it is not clear that there exist no complex interpretations that are associated with visual stability within a given multimedia collection. Take for example the tag "love" on the right image in Fig. 10.1. Upon first consideration, it seems like it would be impossible to generalize over the visual content of images representing "love." However, a second look reveals that the depth of field of this image is quite shallow. The shallowness can be seen because the Golden Retriever is in focus and the background is blurry. When a photographer uses this technique, it gives rise to a feeling of spatial closeness and intimacy. This photographic technique has a visual reflex that creates a commonality between images interpreted as depicting "love." There are a large number of images on the Web that depict "love" in this way. Note we are not arguing that the depth of field is a strong indicator for "love," but rather that it is an indicator that is not initially expected. This example supports the point that it could hold back the progress of research in multimedia content analysis to assume, without further investigation, that complex interpretations never have stable manifestations at the level of signal.

Fourth, it is unclear if researchers are making the right decision to blindly assume the usefulness of principles of compositionality. The Standford Encyclopedia of Philosophy [30] describes the principle of compositionality by stating, "the meaning of a complex expression is fully determined by its structure and the meanings of its constituents—once we fix what the parts mean and how they are put together, we have no more leeway regarding the meaning of the whole." It goes on to state, that this principle is presupposed by most contemporary works in semantics. It is important to note that some work treating "complexity" in multimedia, actually treats compositional combinations of simple concepts, e.g., [18, 35]. Such queries are complex in that they involve multiple parts, but the relationship between the parts is transparent and does not involve complex interpretation. Inherently, there is nothing wrong with compositionality. "Divide and conquer" is a solid strategy that is useful for solving many problems. In its formative stages of the research area, the problem of multimedia content analysis was so formidable, that researchers really had no other choice than to try to break it down, solve individual pieces, and then build up solutions. Now, however, we stand at a moment in which the field has developed

Fig. 10.2 Images illustrating a case in which a complex concept ("wilted flower") might be both more important to users and easier to automatically detect than a simple concept, i.e., "hibiscus" (*left*) versus "tulip" (*right*). *Flickr Credit* Grant MacDonald (*left*) and Mrs. Magic (*right*)

far enough that it is possible to go back and reconsider our dependence of such techniques. Concretely, a noncompositional approach to multimedia means directly targeting the description "wilted flower," for the images in Fig. 10.2, rather than attempting to first detect "flower" and then combine it with the notion of "wilted."

The images in Fig. 10.2 serve to motivate the insight that when researchers focus on detecting simple concepts, they may be forcing themselves to face an unnecessarily difficult challenge. The flower on the left is hibiscus and the one on the right is a tulip. These flowers have a different appearance in these images than they would in the canonical photos typically used to portray these types of flowers. If, however, what is important to the user is that the flower is wilted, and not the identity of the flower, then it makes sense, to skip the simple concept entirely, and jump directly to the complex interpretation "wilted flower." We would like to explicitly point out the relationship between this type of complexity and the work of [3], which proposes an inventory of visual noun pairs based one emotionally colored adjective noun pairs. Here, we do not restrict ourselves to aspects of multimedia with emotional aspects, and also are not specifically interested in adjective noun pairs. However, we mention [3] here, since the idea of moving away from simple concepts to create concepts more closely related to user interpretations is a common underlying theme.

In sum, in the past, simple concepts have served well to advance the field of multimedia content analysis. However, we now have the possibilities at our disposal to address multimedia in terms of complex human interpretations. In the next subsection, we make our notion of complex human interpretations more concrete and provide additional illustrations of why it is imperative that multimedia content analysis takes them into account.

10.2.3 Simple Concepts Versus Complex Interpretations

Two definitions allow us to more formally distinguish a description of multimedia content that involves a simple concept and a complex interpretation. Note that the purpose of these definitions is not to completely characterize "simple concepts" versus "complex interpretations," but rather to provide a diagnostic to help in differentiating them. The purpose of such a diagnostic is to support researchers in identifying

Fig. 10.3 Images described by a simple concept, here, "Phillips screwdriver" (*left*) and a complex interpretation, here, "birdhouse" (*right*). *Flickr credit* loonyhiker

cases in which they might assume that they are dealing with a simple concept, but in which it would be more productive to look at the problem in terms of complex interpretations.

Multimedia description involving a *simple concept*: A description of an image or a video that it does not make sense to question. The validity of the description can be defined with respect to a conventionally accepted external authority.

Multimedia description involving a *complex interpretation*: A description of an image or a video that is acceptable given a particular point of view. The complex interpretation is often accompanied by an explanation of the point of view. It is possible to question the description by offering an alternative explanation. It does not make sense to reference a single, conventionally accepted external authority.

In Fig. 10.3, two images serve to illustrate the difference between a description involving a simple concept and one involving a complex interpretation. The image on the left is described by the *simple concept*, "Phillips screwdriver." Because this image depicts a standard tool, there is also an obvious conventional source of authority that can validate the applicability of the description to the image. In this case, if we would like to identify that authority specifically, it is the toolmaker, Stanley. Note that it is not necessary to actually consult that authority to label this image. The mere existence of an authority that can be readily associated with this image, already establishes that it does not make sense to question whether the image depicts the simple concept or not.

The image on the right is described by the complex interpretation, "birdhouse." We consider "birdhouse" to be a complex interpretation of this image because it is possible to contest its appropriateness as a description of the image. For example, a human looking at the image could plausibly take point of view that because it is shaped like a beehive it should be interpreted as "beehive". Alternately, someone could say that because it is hanging indoors no birds could possibly be living in it and it should be interpreted as "living room accessory."

In sum, in this subsection, we have introduced definitions to formalize the notions of simply concepts and complex interpretations. The strength of the definitions is not in their absolute ability to characterize these notions, but rather in their ability to allow us to distinguish the two. In the next subsection, we turn to examine the limitations of simple concepts in describing multimedia, and the issues that arise when a complex interpretation is unnaturally forced into the form of a simple concept.

10.2.4 The Limitations of Simple Concepts for Describing Multimedia

We illustrate the use of multimedia descriptions involving simple concepts in research on multimedia concept analysis by discussing an example of a multimedia ontology, LSCOM [13]. The ontology has been used widely for multimedia concept detection tasks, e.g., TREC Video Retrieval Evaluation (TRECVid) [28]. These tasks involve automatically assigning semantic tags representing visual or multimodal concepts to segments of video. TRECVid has a track record of having productively exploited simple multimedia concepts in order to advance the state of the art of multimedia content analysis.

Human judges use LSCOM to annotate video by watching the video and assigning the concepts in the ontology to individual video shots. The human annotators annotated the images by making references to definitions formulated to characterize the concepts. For example, "artillery" is defined by the LSCOM task with the sentence, "artillery includes mortars but not tank guns" [13]. This definition represents a simple concept used to describe multimedia according to the definition above. Note that the situations in which "artillery" is used are a priori well-defined. The use of the concept "artillery" is limited to military-related settings and humans who participate in military or combat activities share a common understanding of what "artillery" is used for and what the concept refers to. The ultimate authority on whether an image or video depicts "artillery" lies with someone who has had military training. In other words, the military provides an authoritative definition that can be used to resolve questions of interpretative variation. Note that we do not claim that it is technically simple to automatically determine whether or not an image or a video depicts the simple concept "artillery". Rather, we point out that "artillery" is a simple concept from the point of human understanding.

Next to "artillery" the concept of "dresses" is also included in LSCOM. The concept of "dresses" serves to reveal the limitation of simple concepts. Since styles vary widely across time, region, and social convention, there is no single, obvious authority that can be consulted in order to create a definition of "dresses". The LSCOM definition of "dresses" is "People wearing dresses or gowns Arab men tend to wear them. Also of course women in dresses. The dresses should at least reach the knees" [13]. In contrast to "artillery", it is relatively easy to argue that this definition represents one interpretation of "dresses", but that other interpretations should also be considered admissible. For example, the definition excludes dresses depicted without people, i.e., as they occur in shop windows or certain clothing catalogs. Situations are imaginable, for example, online shopping or personal photo album creation, where users might be surprised by the exclusion of dress-models shorter than knee-length from the definition of "dresses". This example illustrates that if a system forces a complex interpretation to be covered by a simple concept, it can radically fail to capture the way in which an image would be understood or used by a human user.

One tempting approach is to "fix" the simple definition of the concept by extending it each time an exception is encountered. For example, one could append the following sentence to the definition, "…dresses depicted on hangers or on mannequins should also be included." However, this solution proves to be unhelpful with respect to the larger picture. If a concept is to be broadened, it is important not to broaden it arbitrarily, since it will lose its meaning, i.e., its ability to isolate a usefully constrained set of multimedia items from the larger pool. This problem of meaning loss is already reflected in the fact that the definition in LSCOM conflates Western-style women's "dresses" with men's robes in a non-Western setting. Such a conflation is inconsistent with which dictionary definitions of the word "dress" and also obscures the distinction between categories as "thobe" and "galabiya". The definition of "dresses" in LSCOM provides a compelling illustration that simple concepts will not always be appropriate to characterize multimedia in a way that is consistent with the vantage points of the users of a multimedia retrieval system.

Thus far, we have argued for the importance of addressing complexity in human interpretations of multimedia. We have presented definitions that can be used to distinguish "simple concepts" and "complex interpretations" and have discussed the dangers of conflating the two. In the next section, we go on to discuss the nature of complex interpretations in greater depth.

10.3 The Origins of Complex Interpretations

As stated in the introduction, in this article, we advocate that complexity of human interpretations of multimedia should be first recognized, and then the aspects of complexity that are most important to users should be isolated and tackled. As a step to recognizing complexity, in this section we discuss the conditions that give rise to complex interpretations. Then, we discuss the nature of complex interpretations, in particular, relating the discussion of complex interpretations in this article to existing notions familiar from multimedia research, such as "affective impact" and "subjectivity".

10.3.1 Context and Complexity

In this section, we elaborate further on our initial observation that different people assign different interpretations to the same multimedia, and that these differences

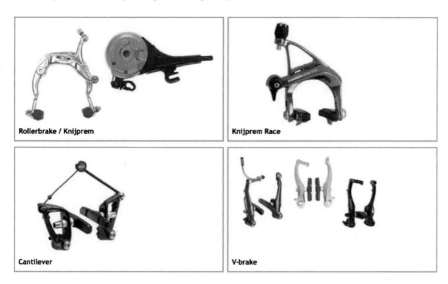

Fig. 10.4 Images used by a bicycle parts website for the purpose of illustrating the different bicycle parts that they sell can be characterized by simple concepts (http://www.fiets-onderdelen-online.nl)

depend on when and where the interpretation is being made. We regard the aspects of when and where as the *context* in which the multimedia is being interpreted.

Hypothetically, we can consider the number of contexts in which multimedia can be interpreted to be infinite. Practically speaking, however, it is useful to narrow contexts down to include only the specific ways in which people frequently or typically use multimedia. Within a context, we can expect the range of interpretations assigned to the same multimedia content by different interpreters to be radically restricted. We base this expectation on the assumption that in a given context, multimedia is used as a form of communication. We do not commit ourselves to particular expectations concerning the strength of the consensus among people using multimedia with a given context. Instead, we simple build from the insight that if everyone assigns a different interpretation to an image within a context, the context will be unable to support human communication and become irrelevant. This insight makes it reasonable for us to assume that a common context must be characterized by a minimum level of consensus.

Let us begin with an example of a context of use: a website selling bicycle parts that uses images to depict the parts, as illustrated by the examples in Fig. 10.4. This context admits very little variation in human interpretation. The images can be described with simple concepts, e.g., "cantilever brake," that identify the items offered for sale by the website.

In the context of this website, it is safe to feel confident in assuming that humans interpret these images to depict the bicycle parts that are being sold. We can exclude, for example, that they are meant to record an interesting experience of the photographer, or demonstrate certain techniques for taking a photograph. Since the images are used within a clearly specified context, multimedia researchers can expect that simple concepts will provide useful descriptions and are a good choice.

Note that, theoretically, it is not impossible that someone would frame an image of the "cantilever brake" from this website and hang it on their wall as art. However, in doing so, this person is not acting according to the purpose of the existence of the shopping website, namely, buying and selling bicycle parts. Instead, the person has taken the image outside of the context of its use. The challenge that faces the multimedia researcher is how to delineate the context. The bike parts website is relatively simple to handle, since it has a well-defined purpose. In general, however, it is important that in advance of developing multimedia content analysis algorithms, the multimedia researcher understands the type of multimedia system in which the algorithms are to be used. The ultimate system for which the algorithms are intended will define the context of use. This context imposes limitations of the interpretations of visual content that automatically generated descriptions of multimedia content should admit.

The rise of social multimedia has also seen the rise of contexts in which images are used for a broad range of purposes. In such situations, what on the surface appears to be a single context (e.g., a photo sharing website) is ultimately more productively understood as a multiplicity of a large number of subcontexts that overlap, evolve, and elude precise delineation. Ignoring or repressing the complexity of possible interpretations in these contexts threatens to cut off the possibility of finding solutions to the underlying problem. For example, an image of a quilt is a picture of a cozy sleeping cover (i.e., something that makes a bed comfortable), but it is also a picture of a craftwork (i.e., something that was made by hand following an age-old tradition).

Note that we are not stating that the image is ambiguous. Note that a quilt neither implies nor necessitates either a bedcover or a craft. Rather, the point that we are making is that when this image is used in the context of a social image-sharing website (i.e., Flickr), it can be equally well be described in either way, depending on the point of view taken. The two descriptions exist legitimately side-by-side within the context. In our view, the key to handling such cases is to consult many human interpreters in order to gather large amounts of information that reflect the range of perspectives that people take on social multimedia online. We do not aim to cover every single possible interpretation, but rather we wish to analyze the interpretations and find the dominant ones. In other words, we identify those interpretations that are most important to users when they are creating, sharing, and consuming the multimedia material. We then add these dominant interpretations to the aspects of multimedia that we design our multimedia content analysis algorithms to cover. Such collection of information is made possible by the use of crowdsourcing platforms, which we will return to discuss in detail later.

The sensitivity of interpretation to context can be illustrated by imagining the image in Fig. 10.5 in another context, for example, on the website of a quilting competition. Here, a description that characterizes the image as depicting a bedcover or a quilt conveys little information. Instead, aspects of the image such as the exact pattern of the blocks in the quilt, the colors used by the quilter, and the quality of the quilting work becomes important. Multimedia content analysis developed for a

Fig. 10.5 Image shared on Flickr, serving as an example of an image best described with a complex interpretation. *Flickr credit* lisaclarke

multimedia system used by a quilting competition website requires entirely different multimedia descriptions.

The challenges faced by researchers are making an informed decision about which context is being addressed, collecting information on the scope of possible human interpretations of multimedia in this context, and deciding which interpretations should be considered dominant, and for this reason most useful to form the basis of multimedia descriptions. Note that this process effectively addresses the problematic example discussed above, in which the concept of "dresses" was expanded to include men's robes. We acknowledge that, technically, it may be useful for a multimedia content analysis system to leverage a "Non-trousers garment" category that conflates men's robes and women's dresses. Visually, this category is presumably distinguished by the presence of a human with no discernible legs. However, the system must also be able to make the distinctions necessary to describe images in a way that reflects that interpretations given them by users in the context of use of the system.

10.3.2 Context Abhors a Vacuum

Context is a challenge for multimedia research, since it is very difficult to measure or control. Researchers are often tempted to approach the problem of generating descriptions of multimedia by first ignoring context, with the idea that it will be added later. However, multimedia content does not arise out of a void, rather is always created with a context. A human looking at, for example, an image necessarily attempts to recreate that context. This attempt is a natural and inherent part of the effort to understand the image. In this subsection, we make the point that it is critical for multimedia researchers to understand that human interpretations may be impacted by context, even in cases where context has been carefully removed, or does not obviously appear to be present.

Fig. 10.6 Two images of cars. These images trigger the human interpreter to reconstruct a context, with great likelihood this context is "car show". *Flickr credit* harry_nl (*Left*) and ptyntofmyld (*Right*)

We express this point by the phrase, "Context abhors a vacuum." When a multimedia content item (i.e., an image or a video) is presented to a human viewer without any indication of where it was taken or what it is used for, the interpretation that the viewer gives to that item will be impacted by the viewer's attempt to construct or reconstruct a context for that image.

We present two different sources of evidence that human interpreters create or assume a context for an image during the process of interpretation. The first source of evidence is practical. Consider the two images of cars in Fig. 10.6.

These images fit with the simple concept description "car." However, unless the human interpreters have been explicitly asked, "Tell me only whether or not the image depicts a car," they will not simply say, "car" and move on. Rather they will note that these cars are not portrayed in motion, that they have no drivers, that they are the central objects in the photo, that they appear to be on display, and that the relevant context for both images is cars being shown at car shows. It is important to realize that human interpreters are not able to look at these images and not also see something in addition to "car." Instead, humans naturally, and largely unconsciously, fill in the blanks in order to create a context.

The second source of evidence that human interpreters create or assume a context comes from *accommodation of presupposition*, a mechanism that humans use to interpret natural language. In the field of linguistics, much formal work has been devoted to understand the mechanics of presupposition. Presupposition in human language is complex, but Beaver and Geurts provide an accessible overview [2]. Here, we present a short, informal explanation of presupposition. Presupposition is information that is not directly asserted by the person who says a sentence. Rather, the truth of the information is taken for granted. For example, in the course of a conversation between two people, one person can suddenly say out of context, "I need to pick my cat up from the vet now." The person is explicitly stating that he needs to go to the vet. He is not explicitly stating that he has a cat. Rather, the sentence takes for granted that he has a cat. His interlocutor may not have known that the speaker has a cat and decide to state, "Oh, I didn't know that you had a cat."

However, critically, it is not necessary for the larger context (i.e., that the person has a cat, that the cat needed to go to the vet) to have been previously discussed or explicitly clarified in order for the sentence, "I need to pick my cat up from the vet now." to be used successfully in a conversation. Rather the speaker can simply reply on presupposition. In other words, he can simply assume that an indirect mention of the cat will trigger the listener to accommodate the presupposition of situation, i.e., construct the context including at least one cat that belongs to the speaker, that was also in need of a vet visit.

Presupposition accommodation happens so quickly and so automatically that it is easy to assume that it is not important to understanding language. In order to more clearly see its importance in human conversation, it is helpful to consider how the same conversation would be held, if presupposition accommodation were not used. The person would say: "By the way, I own several cats. One of these cats was recently in need of a vet trip. I took the cat to the vet and left it there. I need to pick my cat up from the vet now." Here, the reaction of the other person is, "Why are you giving me all the details, when what you really want to say is that you need to leave now because of a previous commitment?"

If humans fill in the context information in this way while interpreting each other's sentences, it is not surprising to find a similar mechanism at work when humans interpret photographs. Indeed, it is not necessary for someone to understand every detail of an image in order to have the feeling that they understand what the image is about. For example, humans will be able to interpret the images in Fig. 10.6 without knowing at exactly which car show they were taken. Instead, they pursue a strategy of filling in the blanks, i.e., constructing or reconstructing the context necessary in order to be able to make sense of the image.

In sum, it is important to be aware of the principle, "Context Abhors a Vacuum." This principle reminds multimedia researchers that context is always a factor in how humans interpret images or videos. Even when there appears to be no context, researchers should still take caution and use a mechanism for context control. In this way, they can exclude the impact of implicit and possibly unexpected assumptions of human interpreters on their interpretation of multimedia context. We will return in Sect. 10.5 to introduce a specific technique for controlling context that can help to reduce the variability in human interpretations of multimedia, without forcing complex interpretations to be represented as simple concepts.

10.3.3 Complex Interpretations Versus Subjective Interpretations

In this subsection, we relate the notion of complex interpretations of multimedia, as it is presented in this chapter, to existing notions of subjective interpretations of multimedia. This discussion is intended to provide further support for researchers in recognizing different aspects of complexity, and in determining which are most important to users of specific multimedia systems.

In the field of multimedia research, "subjective" descriptions of multimedia that are considered to be those descriptions that include users' personal opinions, their judgments of esthetic value, their emotional responses, and their impressions of having a high-quality multimedia experience. Such characterizations of multimedia are referred to as "subjective," since it is the point of view of the "subject," i.e., the specific human user, that is the factor that determines their validity. A good way of identifying a subjective description of multimedia is to ask if it follows the principle behind the maxim, "Beauty is the eye of the beholder." If a user says that a photo is appealing, then it is, by definition, appealing to that user. The user him- or herself is the ultimate authority on subjective aspects of multimedia, no external authorities are involved.

Differentiating subjective interpretations of multimedia from complex interpretations requires careful consideration, since the two can easily be confused. Colloquially, the phrase, "Beauty is in the eye of the beholder." is used to express the fact that we cannot expect people to be in agreement with each other when they make subjective judgments. Indeed, a key challenge facing research in the area of subject aspects of multimedia is the variability of human judgments. When studying subjective aspects of multimedia, it is not possible to expect a high level of consensus among users.

If we as multimedia researchers are to fully understand the richness of human interpretations of multimedia, it is important to reflect carefully on what should be considered "subjective" aspects of multimedia. Although it is in general safe to assume that subjective aspects of multimedia are associated with less-than-perfect consensus among human judges, the converse is not the case. In other words, it should not be assumed that interpretations of multimedia that enjoy less-than-perfect consensus among human judges are necessarily subjective.

The danger of too quickly assuming that an interpretation is subjective is that it leads us to overlook both its stability and the predictable sources of its stability. Since subjective judgments emphasize the point of view of the subject, we generally exclude points of reference beyond the subject when studying subjective aspects. If complex interpretations are considered to be subjective, it is easy to take the attitude, "They're making it all up anyhow." and go back to attempting to express all descriptions of multimedia in terms of simple concepts. Concretely, this danger manifests itself in the reflex of multimedia researchers to discard, a priori, the possibility that complex interpretations could ever be sufficiently consistent at the signal level (e.g., visually invariant enough) to be modeled robustly, or that external information sources could ever be helpful in doing so.

Because of this danger, it is important to keep our understanding of complex interpretations of multimedia focused on human judgments that are objective. For objective descriptions of multimedia, the human user is important, but is not the ultimate point of reference. Instead, "objective" views of multimedia make use of an external reference, such as a definition (e.g., as provided by a dictionary) or a set of examples. In this way, the focus is kept on the "object" and not on the person judging the object. The classical "objective" description of multimedia content is one that involves a simple concept (defined above under "Multimedia description involving a simple concept"). This description makes use of an external reference provided in the form of a conventionally accepted authority. For this type of description, a high-level of consensus among human judges can be expected.

Although it may be less immediately obvious, complex interpretations of multimedia (defined above under "Multimedia description involving a complex interpretation") are also objective in nature. When carrying out complex interpretations, humans make reference to multiple points of view. These points of view cannot be considered inherent to the subject, i.e., they do not depend on the "eye of the beholder." Rather the points of view are associated with external reference frameworks or vantage points. As mentioned in the definition, the points of view are often associated with different explanations that human judges will supply that serve to account for the reasons that they have decided to assign a particular interpretation. Because multiple points of view are possible, it is not reasonable to expect that complex interpretations of multimedia will yield high consensus among human judgment. However, critically, the lack of high consensus does not make complex interpretations subjective judgments. Rather, in the case of complex interpretations, as with simple concepts, it is also the view on the "object" and not the "subject" that is important.

Because we take a practical approach to multimedia, we take the position that "subjectivity" versus "objectivity" should not turn into an lengthy debate, but rather that the distinction should serve to inform the types of multimedia content analysis we develop and the types of multimedia systems that we build. For this reason, we suggest that instead of referring to aspects of multimedia as "subjective" or "objective," researchers should look directly at two factors: first, the level of consensus among human interpreters considering these aspects, and, second, the sources used by interpreters as reference, i.e., whether human interpreters make their judgment, but looking inward to their personal preference, or looking outward to external reference frameworks.

In order to illustrate the way in which we propose that these considerations should be practically applied, we consider the concrete case of the Flickr image shown in Fig. 10.7. In Table 10.1, we give examples of plausible descriptions for this image. The descriptions are chosen so that they illustrate the different levels of consensus and different sources that can be used as reference.

We have included two separate columns Table 10.1 to encode the consensus of human judges concerning the multimedia description. Two columns are necessary because there are two ways for human judges to connect a description with multimedia content. First, they can produce the description unprompted, and, second, they can validate an existing description that has been produced by someone else. We assume that human interpreters are willing to agree to accept more descriptions than they would naturally produce themselves, i.e., they can be expected to validate more descriptions than they personally actually contribute. For these reasons, we include separate predictions of "Production consensus" (the proportion of human judges who volunteer the same description) and "Validation consensus" (the proportion of human judges who accept the description). It is important to note that the

Fig. 10.7 Example image used to illustrate levels of consensus achieved by human interpreters, and the sources of reference the use. *Flickr credit* Scarleth White

consensus values in this table are illustrative and not values that have been actually estimated on the basis of human-contributed information. One of the keys to making the decision concerning which aspects of human interpretation should be considered most important is the ability to make good estimates of these values.

Here, we briefly discuss each of the example descriptions. In the first row, "bird" is an example that fulfills our definition of a description involving simple concept. Unlike the example of "artillery," already discussed above, there is not a specific institution that acts as an external authority on the matter of what should be considered a "bird." Rather, the authority is the established referential practices of the general community.

Formally expressed, "bird" is a basic-level concept. The basic level is a division of the world into categories at the level of abstraction that humans find most natural. The fields of cognitive psychology and linguistics have accumulated a great deal of evidence concerning the existence and the nature of the basic-level concepts [17]. A good working definition of the basic level is that it is the "default" level of abstraction that is most commonly used and is chosen by adults as the most appropriate level to first teach children. Note that we would still consider a description like "seabird" a simple concept according to our definition above ("Multimedia description involving a simple concept"), although it is not a basic-level concept. In this case, the conventional external authority would be a bird expert. Interestingly, the bird expert could probably also supply the genus and species of the particular seabird pictured in Fig. 10.7. If the pool of human judges is taken to be general Internet users, such experts would be quite rare in the pool. (In support of this statement, we note that the authors were unable to determine the correct Latin name for this bird.) We would predict low consensus among people either contributing or validating the species of the bird in this image. Descriptions that can only be produced by experts are classic examples of cases where low consensus is to be expected, but should never be used to conclude that the description should be considered subjective.

In the second through fourth rows, "lovely weather," "unusual seagull," and "freedom" are all complex interpretations. The projected consensus that these descriptions would enjoy among human interpreters is hardly overwhelming. However, these

Table 10.1 Instead of the "Objective versus Subjective" distinction, descriptions of multimedia can be classified using levels of consensus between human judges and the source of reference used as a basis for the description

Example of possible description	Production consensus (projected)	Validation consensus (projected)	External reference for description	Type of description
"Bird"	High	High	Basic-level concept	Simple concept
"Lovely weather"	Medium	High	Preferred weather for photographers.	Complex interpretation
"Unusual seagull"	Medium	Medium	Similar images on Flickr	Complex interpretation
"Freedom"	Low	Medium	Conventional symbolism	Complex interpretation
"Scary"	Low	Low	None	Emotional impact

descriptions are clearly not to be considered subjective, since they all make reference to external reference frameworks. The image in Fig. 10.7 depicts lovely weather from the point of view of someone who wants to make a photo. However, seen from another point of view, e.g., the sky is also blue on a cold day, "lovely weather" might not seem clearly appropriate as a way to describe this photo. The image depicts an "unusual seagull" with reference to other images of similar seabirds on Flickr (which tend to have different characteristics, e.g., yellow and not black beaks.) Note that this description may technically be considered wrong according to the bird expert, who would be able to determine if the photo in fact depicts a tern and not a seagull. However, users commenting on this photo describe it as a gull. The image depicts "freedom" because it evokes the book Jonathan Livingston Seagull, which, in the 1970s, established the image of a seagull in flight as a symbol of striving for freedom from conformity.

Now that we have discussed examples of complex interpretations of this image, we turn to discuss *connotation*, which is a concept frequently occurring in discussions in the literature on how humans interpret images. A key reference on connotation is the work of Roland Barthes, and especially his essay, "The Photographic Message," included in his 1977 book, *Image Music Text* [1]. Barthes characterizes connotation as "the imposition of second meaning on the photographic proper" (p. 20). This "second meaning" is encoded in the photo by choices made by the photographer in terms of content, but also layout, lighting, and other technical choices. Humans can interpret images in terms of connotations because connotations are highly institutionalized, meaning that they are consistently and conventionally used to convey the same message. A classic example is the *nuit américaine* technique used in film, which changes the look of a scene in a conventional way that is used to signal to the audience that the scene should be interpreted as taking place at night. The complex interpretation "freedom" applied to the image in Fig. 10.7 can be interpreted to be a connotation. However, it is important to note that it is not possible to cleanly equate complex descriptions and connotations. The complex interpretation "lovely weather" does not appear to be a secondary meaning imposed on the photograph at all. It is simply something that one person might see directly in the photograph, and someone else, taking a different point of view, might contest. Historically, however, authors discussing connotation have also discussed divisions similar to our distinction between simple concepts and complex interpretation. Rafferty and Hidderly mention some of the major points in a chapter entitled, "Using Semiotics to Analyse Multimedia Objects" in their 2005 book [26].

Finally, in the bottom row of the column, "scary" is a truly subjective description. Here, the human interpreter describes a photo with respect to personal emotional reaction and not with respect to an external reference. The level of consensus is projected to be low. It is important to note that there is only a subtle difference between true subjective descriptions (i.e., ones that fit "Beauty is in the eye of the beholder.") and descriptions that report opinions, esthetic judgments, and emotional reactions that can be related to external influences. For example, it is possible that quite a few people feel that this image is "scary" because it is not uncommon to have had the childhood experience of being terrified by Alfred Hitchcock's film, *The Birds*. Also, the image respects the Rule of Thirds, which is applied by photographers in order to achieve balance in their images. It is not surprising that people in general should find a balanced photo more appealing than an imbalanced one, and should have a greater tendency to describe this photo as beautiful.

These comments on connotation ("freedom") and the subjective description ("scary") support our point that ultimately it may be more productive to focus on how the crowd interprets particular images in particular contexts, rather than being guided by theoretical distinctions. In other words, we believe it is most helpful to analyze descriptions of multimedia with respect to consensus levels and to external references, than insisting on particular definitions of subjectivity and connotation. This empirical, quantitative approach has only recently become possible. It is the rise of crowdsourcing that allows us to obtain sufficient input from enough different human interpreters to make possible useful estimates of consensus.

In sum, up to this point, we have defined descriptions of multimedia involving complex human interpretations, we have argued that they are important for making multimedia systems useful to users, and we have discussed their relationship to other types of descriptions of multimedia. In the next section, we move on to practical techniques for making the first step in taking complex interpretations into account in research on multimedia content analysis.

10.4 Crowd-Based Collection of Multimedia Interpretations

In this section, we turn to discussing crowdsourcing as a tool with which we can gather a wide range of interpretations from human interpreters. Before diving into the details of the discussion, we explicitly recapitulate the underpinnings of our approach. We use crowdsourcing to collect information on possible points of view referenced by humans when they are interpreting multimedia. In particular, we create catalogs of depicted elements and other aspects of multimedia that are important for interpreting multimedia from these points of view. The approach is highly practical in nature. We do *not* advocate creating an exhaustive inventory of all possible points of view, elements, or aspects that could possibly be relevant. Rather, we are interested in effective methods for exploring, free of our own biases, the range of possible interpretations, and then, focusing in on the most important ones. By consulting the crowd, we aim to determine which particular interpretations may be dominant for a certain type of multimedia data, or for a certain application scenarios in which that multimedia data is used.

The intended benefit of this approach is to provide multimedia systems with descriptions of multimedia that more closely fit human expectations and to prevent the issues associated with arbitrarily restricting or broadening the definition of simple concepts. Critically, under our approach we accept that many different legitimate complex human interpretations of multimedia exist side-by-side. We do not force the different human interpretations to be consistent with one another.

10.4.1 Crowdsourcing for Multimedia

Crowdsourcing is the process of producing value or services in a manner that takes advantage of both human intelligence and human multiplicity. What needs to be done and who does it come into connection outside of the locus of a conventional workplace, where specific tasks are delegated to specific agents at specific locations. The original definition of crowdsourcing was formulated by in 2006 by Jeff Howe [12], and has later been refined by many authors, include Quinn and Bederson [25]. Here, we focus our attention on crowdsourcing platforms, i.e., online task markets where crowdworkers can carry out task for taskaskers in exchange for a reward. Typically, the task is quite short, and often referred to, for this reason, as a "microtask." Mechanical Turk[1] is a major crowdsourcing platform run by Amazon. Other examples include CrowdFlower,[2] and Microworkers.[3] Crowdsourcing platforms make it possible to collect input from very large and diverse groups of people. Using such platforms, researchers can ask questions about multimedia items, i.e., images, sound recordings, and videos, and gather a set of responses that reflect how human beings interpret multimedia.

The number of people who can be reached via crowdsourcing platforms is significantly larger than the number of people who could practically participate in a conventional lab-based experiment or field study. Because crowdsourcing makes it possible to access a high volume of feedback from a diverse set of human subjects, it becomes feasible to study the variety and nuances of human interpretations from perspectives that were previously not possible. If we can make use of crowdsourcing to understand the range of possible interpretations for multimedia, it becomes possible to build automatic multimedia indexing systems that better serve the needs of their human users.

Crowdsourcing has been successfully deployed in human behavior research and to run experiments with large subject pools [15, 22]. In the field of multimedia, it has proven very valuable to researchers in the area of Quality of Experience [4, 11]. Quality of Experience studies are closely related to users' perception. For example, they often require crowdworkers to provide answers related directly to sensory information (e.g., "Which image is sharper or brighter?"). Such studies resemble our proposed use of crowdsourcing to gather complex human interpretations since they take advantage of the size and the diversity of the crowd. The difference is that the crowdsourcing studies discussed here involve providing interpretations, a

---

[1] https://www.mturk.com

[2] http://www.crowdflower.com

[3] http://www.microworkers.com

cognitive activity, whereas providing self-reporting on personal experience has a large perceptual component.

Crowdsourcing platforms have been demonstrated to be just as reliable as conventional methods used to gather image labels [21]. However, extending this work to the goal that we pursue here, namely, collecting complex human interpretations of multimedia content, is not a trivial undertaking. The reason is that the tasks most often offered on crowdsourcing platforms are "mechanical" in nature, as reflected in the name "Mechanical Turk". Mechanical tasks are highly routine and usually do not require specialized knowledge. In addition to proving concept labels for images, mechanical tasks in the area of multimedia annotation include tasks such as segmenting objects and matching the contents of images, e.g., faces fall into the category of mechanical tasks. Two example cases [8, 34], with which we have had direct experience, illustrate these types of task. For mechanical tasks, it is most often the case that an external standard exists that can be used to determine whether or not a crowdworker has provided reliable input. Mechanical tasks require little interpretation from the side of the user. A typical mechanical task would ask a user to confirm whether or not the image in Fig. 10.7 shows a "bird." Mechanical tasks monitor the quality of the responses being provided by the crowdworkers by interleaving "validation" questions (i.e., questions for which the correct answers are already known) with the "target" questions (i.e., the questions for which the answers are being gathered). Workers who give incorrect answers to the validation questions are assumed to be providing incorrect answers on the target questions as well, and their work is filtered or rejected.

In order to successfully exploit crowdsourcing for the purpose of collecting complex human interpretations, the key challenge is designing effective tasks. The tasks must be capable of moving beyond collecting responses to "mechanical" questions to eliciting complex user interpretations for images and videos. Because such user interpretations will come from a variety of points of view and are not predictable in advance, it is impossible to judge the quality of a crowdworker's response by comparing it to a standard. In fact, when we use crowdsourcing to collect human interpretations, we are using it exactly because we have no idea of what the responses might be. In contrast to mechanical tasks, the crowd is not helping researchers save time with annotation, rather they are providing researchers with a real-world lens through which to view their research problem.

Even tasks that upon first consideration appear to be mechanical tasks, can involve an unexpected component of interpretation. Such a case arose during the construction of ImageNet. ImageNet is a hierarchical Wordnet-based image database [7]. The concepts depicted by each image have been verified using a crowdsourcing platform. The verification task asked workers to confirm whether images depict certain categories. To perform this task, crowdworkers are referred to a Wikipedia page. The verification method used takes into account that inter-annotator agreement is smaller on some categories than it is on others. In the example given by Deng et al. [7], the verification method automatically collects more input from the crowd before verifying an image as depicting a "Burmese cat" than it does to verify an image as depicting a "cat." This work can be considered to touch on some of the issues that we are addressing here, since it explicitly proposes a methodology to make use of descriptions of multimedia for which there may not be a high level of consensus among human judges.

10.4.2 Trusting the Crowd, not Ourselves

In order to interact with the world around us, we, as humans, must learn to trust our own interpretations. Trusting our own interpretations means not being easily distracted by alternative interpretations that may be offered by others around us. When we, as researchers, use crowdsourcing platforms to collect complex interpretations for multimedia, it is necessary to invert these habits. In other words, it is important that we trust the crowd more than we trust ourselves. The most direct explanation for why this inversion is necessary is that a single person cannot produce diversity of a larger population. However, there is quite a distance between understanding that diversity is necessary, and actively allowing ourselves be guided by other perspectives. Often, even when we think that we are being open to diverse points of view, we are actually still caught in our own world, which has closed back in on us. In this subsection, we cover a couple of potential biases that are impossible for multimedia researchers to avoid. These biases provide concrete reasons why it is essential that we do not take the attitude, "we know better than the crowd." Instead, we should always assume, given that we have designed our crowdsourcing task carefully, that "the crowd knows better than we do."

The first potential bias is our understanding of which images are important. Consider again the images associated with the descriptions "Phillips screwdriver" and "birdhouse" in Fig. 10.3. It is true that these examples have been specifically chosen to provide a clear illustration of the contrast between an image that can be well described with simple concept, and an image whose description involves complex interpretation. However, it is important to resist the temptation of assuming that images like the "Phillips screwdriver" image are "common," and that image like the "birdhouse" image are "uncommon." Although the style of this particular birdhouse may make it a rare object, overall, images of rare objects are not necessarily themselves rare objects. Although any given type of object maybe only rarely represented, in a social media collection, rare images compete with common images. If anything, images that are best described by complex interpretations dominate images that are best described by simple concepts on Flickr. We may be attempted to view images like the "birdhouse" image in Fig. 10.3. as noise, or discard them as "improbable" exceptions. However, if we limit ourselves to images that can be described by simple concepts, we are effectively "throwing the baby out with the bathwater." In other words, we are falling far short of solving the overall problem of generating descriptions of social multimedia that are maximally useful to users. Instead, we should leave it up to the crowd to decide which images are part of the signal and which images it is not worth describing.

The second potential bias is our tendency to privilege our own interpretations. When we as multimedia researchers look at a given a photo or video, we have easy access to our own point of view. On the other hand, it takes much more effort to find another human being and ask them for their interpretation. If we ask another person, it is likely to be someone who sits in the next office, whose opinion is highly likely to be correlated with our own due to the close social and intellectual contact. Although it may be possible for us to counteract our own laziness in order to avoid privileging our own points of view, there are other biases over which we have not control. The reality of such biases is demonstrated by the existence of the so-called Bayesian Truth Serum [24]. This effect refers to the tendency of people who are reporting their own opinion truthfully to overestimate the number of other people in the population who share their opinion. Apparently, we are programmed to be overly optimistic about the number of people that we predict to agree with us. For this reason, it is necessary that we, as multimedia researchers, learn to ignore even a very strong feeling of being "right," and instead leave the job of providing interpretations of multimedia content to the crowd.

In sum, this section has introduced crowdsourcing and its usefulness for studying the ways in which people describe multimedia. We have argued that the biases of our perspectives as researchers mean that crowdsourcing can be a source of information that we cannot access in any other way.

10.5 Crowdsourcing Complex Human Interpretations

In this section, we discuss techniques that make it possible to collect complex interpretations of multimedia content from the crowd. In the Sect. 10.5.1, we discuss an example of a task run on Mechanical Turk that was designed to broaden our understanding how humans interpret images in the domain of fashion. Sections 10.5.2 and 10.5.3 moves from the concrete example to present practices that are helpful for guiding the development of tasks that elicit complex interpretations of multimedia from the crowd. Section 10.5.4 presents background information about the development of our approach. Finally, Sect. 10.5.5 presents another concrete example that illustrates the limitations of what the crowd is able to offer.

10.5.1 Example Crowdsourcing Task

In this subsection we present a task that we designed and carried out on Amazon Mechanical Turk. The task had the function of eliciting interpretations from the crowd concerning user-contributed fashion images from an online image-sharing platform (Flickr). It was designed in the context of a project developing multimedia content analysis technology for use in an analysis system that would process social multimedia in order to detect trends in popular fashion. During the process of deciding

Fig. 10.8 Images used in a task designed to elicit factors that influence users when forming judgments on fashion images. *Flickr credit* epSos.de (*left*) and Becca (*right*)

which sorts of categories the multimedia content analysis should focus on, we found ourselves "fixated" on simple concepts corresponding to items of clothing, such as "shirts" and "trousers." We were bothered by the suspicion that users using our fashion trend analysis system would be interested in many other aspects of fashion beyond basic clothing categories. However, we also felt that we ourselves were unqualified to decide which aspects of fashion photos users consider when they interpret them. In order to make a collection of the factors important for users when considering fashion images, we designed a crowdsourcing task. The task collected information from workers both on basic concepts depicted in the images, as well on complex interpretations. In our presentation of the task here, we focus on the results of the task that were relevant for complex interpretations.

The task presented crowdworkers with a fashion image and asked the question, "Do you like what is depicted in the picture?" (possible answers were "Yes," "No," "I am not sure"). Example images are shown in Fig. 10.8. This question was followed by the key question "What do you like/dislike about it?", which required a free text answer. Note that we do not ask these questions because we are directly interested in whether or not the human judges like the images. Rather the aim of the questions is to collect a list of the aspects of the images that human interpreters refer to when they make a decision concerning the images. These aspects correspond to factors that are important for users when judging images in the fashion domain.

Recall from the discussion in Sect. 10.3.1, that complex interpretations humans give to images are related to the context. In order to narrow the input that we received from the crowd to those factors most important for our domain, we needed to focus our task on the context of our application, namely, popular fashion. A critical part of task design is narrowing down the context in which crowdworkers are interpreting the multimedia content.

As a preliminary remark, we point out that, a priori, the popular fashion domain is a good fit with the demographic of workers on Mechanical Turk. Clothing is a universal phenomenon, and we expected that the crowdworkers would have at least a basic sense of what is important when making decisions about what kind of clothing to buy. Note that if our domain were "high fashion" rather than "popular fashion," we would not have the same reason to assume that the crowdworkers on Mechanical Turk would be a good population from which to be eliciting complex interpretations. Finally, the ultimate fashion trend analysis system is intended to support clothing marketers in understanding and opening broader markets. The markets would presumably use the same channel as the one that connects the crowdworkers to Mechanical Turk, namely the Internet, in order to carry out their marketing. We also point out that we do not require a perfect approximation of the target demographic. Our goal was to broaden our understanding of which dimensions are important to people interpreting fashion, and not to build a fine grained model of how fashion judgments are made.

Our main strategy for controlling the context in which the crowdworkers interpreted the images was driven by a simple realization. We found that if we simply presented images to people and asked them, "What do you see in this image?" they may not understand that we are interested in the fashion dimensions of the images, and not other aspects (for example, the people wearing the clothing or the surroundings in which the images were taken). Also, the danger would be high that they would respond with a highly personal opinion, rather than identifying factors that would be relevant for a broader public. Another danger was that the crowdworkers would understand our questions as "searching" for particular answers. For example, they might assume that we are representatives of a certain brand of fashion and are more likely to approve their answers (i.e., accept and pay for their work) if they create replies to be particularly positive or oriented towards our own products. Giving the crowdworkers specific directions, "Do not let your answers be influenced by any particular designs or products" would be counterproductive. Mentioning specific designs might further convince them that we are "searching" for certain answers (although we do not want to admit it). Also, asking people to produce a natural answer to the question, while at the same time unnaturally *not* doing something makes the task more difficult and tiresome and less engaging or fun for the workers.

For this reason, we decided to present the crowdworkers with an interesting backstory and ask them to answer the questions from the perspective of the person in the story. The story also helped to prepare them for the very diverse images that they would see, countering the expectation that they would see professional fashion photographs taken, for example, for a magazine. The text of the story read, "For this task we would like you to imagine you are a fashion blogger. For your blog, you usually collect pictures you find on the Internet. The pictures you collect can be very diverse: they might contain professional models, regular people, fashion from other cultures, or different époques. For you it is not important that the pictures look professional, but that they contain any kind of outfit or fashion-related item such as clothes, fabrics, and accessories. You might also collect pictures of outfits and fashion-related items which you do not like and discuss them with your readers." This text was presented to the crowdworkers at the top of the task before the image.

Fig. 10.9 The affinity diagram used to cluster crowdworker responses into categories each containing a different type of interpretations

The task collected a total of 216 interpretations (three separate interpretations for each of 72 images). The task was carried out on Amazon Mechanical Turk by a total of 50 crowdworkers. Of these, 72 % reported themselves to be female and some 28 % male. The mean reported age was 32 years (Standard Deviation = 7.6). The age of the oldest contributor was 53 years and the youngest was 18 years old. Contributors came from 10 countries. Most of the contributors came from India (41 %), followed by USA (40 %), and Canada (4 %). Chile, Serbia, Croatia, Dominican Republic, Philippines, Poland, and UK had one representative contributor. Contributors from the USA were most active, providing 50 % of the total interpretations, followed by India with 34 %. A total of six responses from five different crowdworkers were discarded because they were not intelligible.

We manually inspected the 216 interpretations and clustered them into classes according to the types of factors to which the crowdworkers made reference. To perform the clustering we used an affinity diagram, shown in the image in Fig. 10.9. Such diagrams are commonly used to sort the ideas arising in a brainstorming session into groups that reflect their natural relationships. Affinity diagrams are created by, first, making individual cards or slips of paper for each factor. Then, slips with related ideas are used to seed groups. The slips are sorted until every slip belongs to a group. The groups are usually reorganized during the sorting process such that the similarity among slips belonging to one group is larger that the similarity between any two slips across groups.

The classes of factors contributing to crowdworker interpretation of social fashion images that we discovered by using this crowdsourcing task are summarized in Table 10.2. The results in the table make it clear that we have broadened our understanding of the fashion domain substantially beyond our initial assumption that the aspects of images that matter most to users are the identity of basic fashion items, such as "shirts" and "trousers." Here, we have identified eight categories of different kinds of interpretations. Note that we did not intend to take all the information discovered in this task into account when making decisions about which kinds of

Table 10.2 Before turning to crowdsourcing, our multimedia content analysis focused on the identity of basic fashion items

Classes of factors contributing to the interpretation of fashion	Input collected on the crowdsourcing platform
Visual appearance of fashion item	Color ("I don't like red sunglasses") and pattern ("the pattern is a little crazy")
Visual appearance connected with a physical property	Cut ("the shoulders are a bit too much") and fabric/material ("princess feel of the gown")
Physical properties related to wearing the item	"Warm," "comfortable"
Aspects differentiating this item from other similar items	"Trendy ties," "unique skirts,"
Aspects reflecting the sort of impact made by the item	"Attractive gloves," "classy purse"
Aspects involving use of the item	"A set of nice business attire for office work and interviews," "The black leather outfit is a great look for clubbing"
Properties of item combinations	"Fun," "whimsy," "sloppy," "not flattering," "Different colors don't look good," "the way of outfitting the pendant is awesome"
Properties of the image	"It's a really unflattering picture," "The picture surroundings are not good"

Crowdsourcing revealed a much broader set of dimensions to which users are sensitive when interpreting fashion images

multimedia content analysis to pursue, rather we intended to exploit the most useful ones.

Particularly useful points that we have observed is that people take color and patterns into account when interpreting fashion images. These points are helpful because the color and patterns of clothes are characteristics that are feasible to automatically recognize (as opposed to "fit" or "style," which is much more challenging). The complexity here lies in the reference to a "crazy pattern." If we had not done this study, we would not have independently arrived at the idea that "crazy pattern" is a worthwhile aspect of fashion. With this knowledge, we can investigate further in order to determine if what people consider to be "crazy patterns" have the sort of stability necessary to train a visual classifier. Not everyone needs to agree on what constitutes "crazy" with respect to a pattern, there just needs to be enough consensus to make this a useful label for use in a multimedia system.

We also found that users are sensitive not just to items individually, but to combinations of items, i.e., the overall "look." Finally, we discovered that overall properties of the image going beyond just the clothes are important to users who are judging fashion. If the picture is in and of itself not technically satisfying photo, this interferes with the interpretations that people give to the fashion it depicts. If we would focus our multimedia content analysis research exclusively on the identities of individual items, we would miss out information on how the images are globally interpreted.

It is important to note that relative to what is possible, this task was quite small in scale, and not particularly sophisticated, since it was designed without an intensive piloting stage. However, the task allowed us in the course of approximately 2 days to significantly improve our understanding of the complex interpretations that people give to fashion images. We expect to gain even more insight with larger, more sophisticated tasks in the future. However, we do note that because of the affinity clustering step, task size is effectively limited to how many interpretations we can manually process afterward.

We conclude this subsection with a couple of additional points on the design of the task. First, we note that the task appears to have attracted a particular population of workers. They are slightly younger that the average worker age reported in the literature, and the ratio of women to men is higher (although women do outnumber men among crowdworkers in the USA) [27]. We conjecture that including the backstory and asking the crowdworkers to answer the question from the perspective of the person in the backstory has a tendency to draw workers to the task who identify with the person. This effect could be helpful in targeting particular demographics for particular multimedia systems. Second, this task has been successful in gathering information from the crowd that goes above and beyond "mechanical" questions. In Sect. 10.4.1, we mentioned that it is difficult to control the quality if a task is not mechanical, i.e., there are no right or wrong answers. Filtering serious from nonserious workers requires looking at workers' work and estimating the level of seriousness and engagement with which they approached the task. Creating a backstory for the task and asking open text answers that invited creativity proved to serve us well as a filter. In the end, less than 3% of our responses were unserious. We conjecture that the backstory and the open-ended questions signaled to the crowdworkers that we as taskaskers would review their work by hand. Apparently, the form of the task communicated to the crowdworkers that we value their work and inspired them to engage seriously with the images and provide us with useful interpretations.

10.5.2 Iterative Task Design

In this section, we move from the examination of a specific example in the previous section to discuss general strategies and techniques that are useful when designing a task that elicits complex interpretations of multimedia from the crowd. Our goal is to design a task with a backstory, which is understandable, appealing and engaging for crowdworkers, who understand that they should project themselves into the backstory, and are able to maintain their motivation and respond genuinely and seriously to the questions posed to them by the task. It is impossible to predict in advance what the crowdworkers, who can be considered users or "the audience" of the task, will find appealing. Because the task has a backstory, in many ways it can be compared to a short story, a film or even a video game. These communication forms are notoriously difficult to design, and film or game studios will invest large amounts of effort into piloting before producing a final product. These considerations lead us to recommend an iterative approach to designing crowdsourcing tasks. In practice, we have found it productive to include three iterations in the design process that precedes launching the task in its final form on the crowdsourcing platform:

- **Initial design** the task is designed and discussed by the researcher and the project team until they agree that it meets the requirements.
- **Internal test** the draft version of the task is tested with a pilot test group of approximately ten people, who can be interviewed onsite. The testers should meet two requirements: they should be able to empathize with crowd workers and they should be able to reflect on and provide feedback about the task. The "sandbox" of Amazon Mechanical Turk is, for example, handy for internal testing.
- **External test** the task is published at a small scale on the crowdsourcing platform. The final question of the task is a box asking crowdworkers to give their opinion of the task and suggestions for improvement.

Because appeal and engagement are critical, the user experience of the crowdworker while carrying out the task is critical. For this reason, we recommend that during the process of task design, and especially internal testing, user experience evaluation should be carried out to detect serious usability issues. Particular attention should be paid to the classical usability dimensions of task efficiency, learnability, and error prevention [19]. Furthermore, the workers' motivation for the crowdsourcing tasks is considered a critical success factor.

To evaluate the user experience, including the crowdworkers' motivation, a wide range of usability inspection methods are available that can involve either target users or user experience experts. A widely used example of the former is the combination of think-aloud and observation. Participants verbalize every thought that comes to their minds, assuming that the verbalizations reflect the contents of their short-term memory [31]. The combination with observation provides a detailed insight into the user's behavior. Testing with, for instance, five to ten people (e.g., students) is sufficient to detect over 80 % of the problems [20]. If done efficiently, this would require approximately only one day of work for preparation, testing, and analysis.

An example of user experience evaluation with experts is heuristic evaluation in which two or more experts evaluate a system using a standardized set of guidelines [32]. Even though it is an efficient way to detect major issues, experts have proven to be unsuccessful in predicting the problems that real-world users encounter [5]. For this reason, we recommend iterative testing, as outlined above, rather than exclusive reliance on experts.

10.5.3 Iterative Elicitation

In the previous subsection, we have emphasized the importance of iterative tests in order to arrive at a well-designed effective task. In this section, we make the point that, ultimately, it may require multiple tasks in order to build a picture of complex human understanding of multimedia that is truly useful. We start our discussion with the observation that crowdsourcing tasks that collect information about multimedia that could not be otherwise inferred or predicted are in fact a type of qualitative research. It is natural, that we turn to established principles that have proven useful in qualitative research to inform our design of this type of tasks. In particular, we propose making use of Grounded Theory, a technique used to inductively build a theory through the collection and analysis of qualitative data (e.g., data from interviews, or observations). The "Theoretical Sampling" approach is a key element in Grounded Theory research, and is defined as, "the process of data collection for generating theory whereby the analyst jointly collects, codes, and analyses his data and decides what data to collect next and where to find them, in order to develop his theory as it emerges" ([10], p. 45). In the case of crowdsourcing tasks that elicit information about complex interpretations of multimedia, the "theory" being described is the set of factors that are important to people when interpreting multimedia in a given domain.

One important point that Grounded Theory has to contribute is that eliciting information from crowdworkers about multimedia should necessarily be a multistage process. In other words, iterations of tasks, one building on the next, are necessary in order to arrive at a maximally informative overview of the aspects of multimedia that are most relevant for human interpretation. As in the example above, the ultimate product is a set of categories or a typology. For tasks in the first iteration, a subset of the dataset or even a manually constructed dataset can be used. If the tasks are manually constructed, careful attention should be paid to generalizability: to what extent do the multimedia content that the crowdworkers are exposed to in the task represent the dataset as a whole? How and to what extent does the manual selection of items from the dataset influence the results? Additional iterations can be used to refine the way in which the task is described to the crowdworkers or to ask the crowdworkers themselves to help in broadening the multimedia data set used to represent the domain. At the end of the process, iterations can be used to confirm and validate the typology and gain insight into its coverage. In Grounded Theory the process of gathering information is carried out by writing codes. Codes labels designate concepts, and, in later iterations, the attributes of these concepts and their interrelations. During the course of iteration, the codes are combined with direct feedback from the workers, leading to refinement of the categories. The cycle of task design, data collection, and coding stops when, it achieves what is referred to in grounded theory as theoretical saturation, i.e., the point at which no additional properties and dimensions for each category can be obtained.

10.5.4 Controlling Context by Creating Context

Our approach to designing tasks for eliciting complex interpretations derives not only from our experience of with the task described in Sect. 10.5.1, but also from experiences that have been reported elsewhere in the literature. The work reported in [33] develops the idea of crowdsourcing tasks with a high "imaginative load." Loosely analogous to *cognitive load*, the burden placed on a person's working memory, the *imaginative load* refers to the burden of projection, i.e., thinking oneself into someone else's shoes and answering a question from their perspective. This work successfully makes use of a crowdsourcing task with a backstory in order collect users' perceptions concerning the similarities of multimedia items.

The work reported in [6] uses not one but several backstories. It presents the crowdsourcing workers with a set of different roles (e.g., blogger, journalist, photographer) and asks them to choose a role and answer the questions in the task from the perspective of that person. The purpose of the task is to collect people's perceptions of whether or not they feel deceived when they see an image that has been digitally manipulated used in a particular context. When they choose a role, the crowdworker is also choosing a context. Practically speaking, the role can also put the crowdworker at ease, allowing them to report their feeling that someone else might accept a manipulated image as not being deceptive, without committing to the position that they themselves would not find it deceptive.

In both of these cases, the backstory has provided a context in which the crowdworker is answering the questions. The presence of the backstory helps to trigger the crowdworker to thing about particular uses to which multimedia is put. Upon first consideration, one may wonder why a crowdworker would pay attention to the backstory since it seems like extra effort. However, if human interpretation of multimedia involves the creation or the assumption of a context, then a task with a backstory actually represents less effort. The crowdworker does not have to invent the content from scratch. We conjecture that the backstory also gives the crowdworker more insight into the reasons why the questions are being asked, and with this more confidence that the answers will be accepted by the taskasker. Ideally the backstory also serves to deter unserious workers, because it is obviously asking for answers that the taskasker will screen by hand.

10.5.5 Within Reach Versus Beyond Control

We have argued in this article that crowdsourcing is an important tool for understanding human interpretations of multimedia. However, it cannot be considered to be the perfect solution to collecting information about factors contributing to complex interpretations of multimedia. For certain aspects of multimedia, it is simply very difficult to elicit information from the crowd. In this subsection, we discuss our experiences with a second crowdsourcing task. The task was a success, and was productive in collecting a large amount of high-quality information. However, during the design process and while analyzing the worker responses, we noticed that there were several examples that served to reveal limitations of the possibilities for collecting interpretations from the crowd.

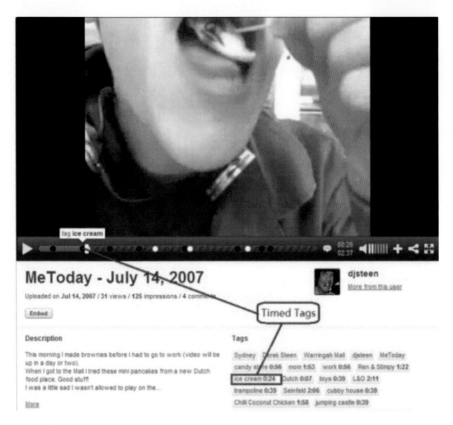

Fig. 10.10 This diagram was included in the task to illustrate the principle of timed-tags to the crowdworkers

The task was designed and carried out in the context of a project studying a special form of video tagging. On some video-sharing websites, such as Viddler.com, tags can be assigned to a time-point by stopping the video and clicking on the navigation bar (Fig. 10.10). We refer to these tags as timed-tags. The goal of the research was to better understand the properties of these tags. Specifically, we were interested in investigating three factors: first, if the tags referred to objects that were visually depicted in the video, or other aspects of the video, second, the way in which the relevance of tags relates to the video in the neighborhood of the tagged time-points (i.e., size of the window or relevance, and the relative position of the tag within this window), and, third, the specific reason for which a person would want to use a tag to start watching the video at this time-point.

From Viddler.com, we collected 5,194 videos with at least one timed tag. On average, there were 3.4 timed-tags for each video. Guided by our research questions, we developed a pilot task using 100 videos and carried out several rounds of testing and refinement. Through the pilot task, we discovered some interesting observations, problems that could be fixed, and issues that were beyond our control.

> **Watch Video Snippets and Tell Us about the Tags**
>
> **Introduction:**
>
> We are carrying out non-profit research at a university to better understand why people tag specific time-points in videos. We refer to the tags that people add to videos at specific time-points as "timed tags". We ask you to watch parts of the video around the tag and answer two questions about how the video is related to these parts. For the final question, we would like you to imagine the tag is not there, and complete the sentence.
>
> You can either use the link to watch the video (Example: Viddler User Interface) or watch it in the lower window. The timed tags are shown on the seeking bar as black points, and they can be seen when you move your cursor to a point. You can also link to the time-point by clicking the hyperlink of the tag.

Fig. 10.11 Introduction of the final version of the task used to collect information about timed-tags related to videos

We start by mentioning the problems that could be straightforwardly addressed. During piloting, when a task does not yield the desired results, a first reaction might be to blame the crowdworkers instead of the task design. However, it pays to remember that the likelihood is high that the task is at fault. Relatively small refinements can be a difference. Specifically, the language used by a task must be correct, clear, and very simple. In our experience, our first idea of simple is often not simple enough. For example, the first pilot we ran on Mechanical Turk had the title, "Watch small bit of internet videos and choose the explanation of video tags." We initially thought that the language of that title was lucid and appealing. However, the task was more successful after we gave it the very short, simple title, shown in Fig. 10.11.

We also find that it was very important to keep the task short overall. Independently of the reward received per task, crowdworkers find work more appealing when the tasks are short. We conjecture it is more pleasant to carry out tasks in series when it is possible remember the questions from one task to the next. In this way, crowdworkers can get into the "flow" or the "groove" of doing tasks without having to re-understand each question the task asks them to answer for each new video. In the final version of our task, it took crowdworkers an average of two minutes to complete the task.

It is also important to keep the tasks of equal length. We originally piloted a task where we would ask questions about up to three tags per video. From the crowdworkers' point of view, it does not make sense that, for the same reward, sometimes they are asked to answer three sets of questions and sometimes one. The uptake by crowdworkers of the final version of the task, which only asked one set of questions about a single tag-video pair, was significantly faster than the uptake of the initial version.

Finally, we noticed the importance of illustrations and examples, such as in Fig. 10.10, and also other forms of "help." The open text question about why someone would watch the video starting at a tag worked best when formulated as a "complete the sentence" question, i.e., "A person watching the video at this time-point would be someone who wants to..." The "complete the sentence" question also was preferred to a multiple-choice question with too many options. Crowdworkers preferred to respond by writing text than to choose one of seven options. These points also helped to improve the uptake of our task among crowdworkers.

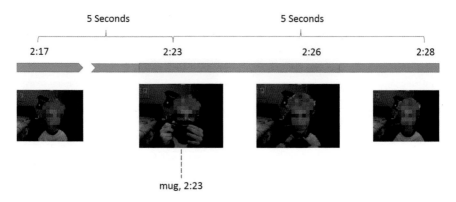

Fig. 10.12 An example in which it was difficult for crowdworkers to separate what they heard from what they saw. The mug is a subject of discussion throughout a longer segment of the video (*blue*), it is visible in the frame for a shorter subsegment (*orange*). The mug was tagged at 2:23. Workers found the tag to apply visually 5 s earlier and 5 s later, although the mug is discussed and not shown in these regions

Other issues that we encountered could not be solved. In particular, we found evidence of the effect that people perceive video as a whole, and have a very difficult time separating the information conveyed by the visual channel and the information conveyed by the audio channel. Recall that in this task, we are particularly interested in the tags that are related to the visual channel, and we would like to know which part of the video is related to the tag. To this end, we used the following sentence, "The next question (question 2) asks you about the connection between the tag and what can be seen in the video. When you answer this question please turn off the sound of your computer and pay attention only to what you see in the video. (If you feel that you can reliably ignore what you are hearing, it is OK if you simply pretend the sound is turned off rather than turning it off.)"

We were pleased when we initially formulated this sentence because we thought that it gave very clear instructions. However, it is obvious that some crowdworkers did not ignore the sound, although we tried to nudge them to think specifically about their own hearing. Figure 10.12 shows an example of such a case.

The example is a video about a man showing his birthday gifts. He talked about a mug for about 30 s, but only showed it in front of the camera for less than 5 s. The tag "mug" was labeled on the time-point where the mug is shown. For this task, two of the three workers who provided information on this tag-video pair thought the tag could be moved 5 s earlier or later and it would still be related to what was seen in the video.

Ultimately, we addressed this issue to the best of our ability by making the entire task more clearly focused on the idea of "tell us what you see when you turn off the sound." We decided that the first question of our original pilot task was distracting. This question asked the crowdworkers a question concerning the tag as a lexical word and not the tag as something that could be seen in the video. In the final version of the task, we eliminated this question. Because we were interested in developing an effective task and not studying how human perception per se, we did not stop to consider the nature of this distraction. We mention here, that what we characterize as "distraction" could be anything from crowdworkers feeling uncomfortable with a task that they feel is drifting "off topic" to a bona fide priming effect. Priming is a memory effect beyond conscious control that causes our response to any given stimulus to be impacted by the immediately preceding stimulus. Although we cannot assert with certainty that we observed priming, we note that priming has been demonstrated to have an impact on crowdworkers' performance in affective crowdsourcing tasks, and is thought to have a role to play in other types of crowdsourcing as well [16].

Finally, we note that a task might be difficult for crowdworkers because of variations in their background knowledge. Conventionally, a differentiation is made between expert and general public crowdsoucing, as in [34]. However, expertise necessary for interpreting social multimedia cannot be equated with encyclopedic or textbook knowledge. For example, in our video collection, the tag "rattail" was assigned to a video in which girls were demonstrating makeup. If the crowdworkers do not realize that 'rattail' is the name of the girl's hairstyle, they may think this tag is not related to the video at all. A general strategy that suggests itself is to ask as many crowdworkers as possible. "Rattails" are relatively rare, but not an entirely obscure classical hairstyle known throughout the world.

In sum, we have used this example to argue that it is important not to assume that unexpected responses of the crowdworkers to a task are the fault of the crowdworkers. Instead, the task must be carefully piloted. However, crowdsourcing does have its limitations and it is important to be aware of them when using it as a tool for eliciting information on complex human interpretations of multimedia.

10.6 Conclusion and Outlook

Multimedia content analysis researchers find complexity in human interpretations of multimedia nearly unbearably frustrating. If there is no single "right answer" to the question of whether a specific image depicts a certain object or a certain video shows a certain event, then how is it possible to develop an algorithm that will generate a useful description of that image or video automatically? Effectively, the complexity arising on the "human side" of the semantic gap creates an impasse threatening to block the progress of multimedia algorithms that are able to generate human-like descriptions of multimedia. For this reason, focusing on diverse human interpretations of multimedia is an important undertaking, although it initially appears to be prohibitively difficult. This article has pursued the modest aim of taking a few practical steps in the direction of a larger solution, although the solution itself lies, presumably, many years in the future.

The approach that we advocate involves two phases, first recognizing complexity, and then isolating the most important aspects of complexity so that they can be tackled by multimedia content analysis research. We have taken the position that the variability of objective characterizations of multimedia should not be considered to be "aberrations" or "subjectivity" in the way that people describe multimedia. Rather, variability in interpretation arises through variability in context of use and is an essential part of how humans make sense of multimedia. As such, we argue, multimedia researchers should strive to addresses the variability of user interpretations of multimedia content in their full richness and complexity. We have discussed reasons for which the interpretations that we, as multimedia researchers, give to multimedia do not reflect those that are assigned by users using multimedia in real-world contexts. In other words, it is not possible for us to expect to "see with our own eyes" the factors that play the most important role for users. Instead, the factors (e.g., visual concepts and other properties of multimedia) that determine how humans carry out complex interpretation of multimedia are something that we must actively go in search of.

Looking forward to the future, it is instructive to compare the approach that we have presented here, with the recent work in the area of computer vision that has made use of crowdsourcing to elicit attributes describing images from humans. Attributes are defined as properties of images that mediate between low level features and high-level categories. In [23], an inventory of attributes is created that are detectable by visual algorithms and are also nameable by humans. The inventory is created so that it is visually discriminative with respect to high-level categories. This work is similar to our own in that it uses crowdsourcing to consult a large number of people concerning their interpretations of images, rather than using a hand-crafted inventory created by a few people. However, the differences are more informative than this similarity.

First, the work in [23] focuses discovering attributes that describe high-level categories that are basic concepts: outdoor scenes and animals. In other words, the ultimate target categories of interest to the user are assumed and not elicited. Our work targets ultimate categories which are of themselves complex.

Second, the Amazon Mechanical Turk task used in [23] does not attempt to create a context for the crowdworkers. The subjects are shown images and asked to "name properties." Initially, this seems like the most straightforward way of asking about images. However, it is actually not straightforward. If the crowdworkers have no clue of context, they need to invent their own. The natural human way to approach the situation is to build a new context: how I describe images in the context of a task that is asking me about the "properties" of images.

The example that [23] provides of a distinction that is considered "unnamable" by crowdworkers is the split between "elephant," "lion," "polar-bear," "sheep" on the one hand and "gorilla," "giraffe," "giant-panda" on the other hand. This distinction is visually discriminative, but crowdworkers cannot name it. We conjecture that if the task had been couched in a context related more directly to a real-world relationship to animals (e.g., selecting images for kid's school projects on the animal kingdom), then people would more easily trigger that "gorilla," "giraffe", "giant-panda" all fall into a "eats shoots and leaves" type of animal, that is often pictured feeding against a foliage backdrop. We do not claim that there is something inherently wrong with the procedure in [23]. To the contrary, the information that it elicits from the crowd is extremely useful and clearly advancing the state of the art. Our point is that the formulation of the task limits the information that can be collected. In our view, the focus on "properties" of pictures that have been isolated from their context of use stands in the way of allowing crowdworkers to connect with the full richness of the complex interpretations that they give to images.

Other work in the area of attribute, such as [9], also uses a task that gives no context. This work mentions the concern of the expense of collecting labels from crowdsourcing platform, and the need to simplify decisions to control expense. However, limited budget should inspire us to think more carefully about whether we are asking the right questions on crowdsourcing platforms, rather than making assumptions to build constraints into the way we ask people about multimedia.

If we are sure that we are focused on what is essential we can avoid wasting resources attempting to solve versions of the multimedia content analysis problem that do not directly contribute to building multimedia systems that support users. Crowdsourcing platforms make it easier than ever before for multimedia researchers should direct their gaze outwards and explore complex human interpretations of multimedia. Ultimately, significant progress in this line of investigation will require a substantial investment of resources, time, and effort. The alternative is that the multimedia systems of the future fail to address what is truly important to human users, and it is those users who are forced to adapt to the system instead.

Acknowledgments The research leading to these results has received funding from the European Commission's 7th Framework Programme under grant agreements No. 287704 (CUbRIK) and No. 610594 (CrowdRec). It has also been supported by the Dutch national program COMMIT.

References

1. Barthes R (1977) Image music text. Hill and Wang, New York
2. Beaver DI, Geurts B (2013) Presupposition. In: Zalta EN (ed) The Stanford encyclopedia of philosophy (Summer 2011 Edition). http://www.plato.stanford.edu/archives/fall2013/entries/presupposition/
3. Borth D, Ji R, Chen T, Breuel T, Chang S-F (2013) Large-scale visual sentiment ontology and detectors using adjective noun pairs. In: Proceedings of the 21st ACM international conference on multimedia (MM '13). ACM, New York, pp 223–232
4. Chen K-T, Chang C-J, Wu C-C, Chang Y-C, Lei C-L (2010) Quadrant of euphoria: a crowdsourcing platform for QoE assessment. IEEE Netw 24(2):28–35
5. Cockton G, Woolrich A (2001) Understanding inspection methods: lessons from an assessment of heuristic evaluation. In: Blandford A, Vanderdonckt J, Gray Ph (eds) People and computers XV–interaction without Frontiers, pp 171–191
6. Conotter V, Dang-Nguyen D-T, Boato G, Menéndez M, Larson M (2014) Assessing the impact of image manipulation on users' perceptions of deception. In: Proc. SPIE 9014, Human Vision and Electronic Imaging XIX, 90140Y
7. Deng J, Dong W, Socher R, Li L-J, Li K, Li F-F (2009) ImageNet: a large-scale hierarchical image database. In: CVPR'09: IEEE conference on computer vision and pattern recognition, pp 248–255
8. Galli L, Fraternali P, Martinenghi D, Novak J (2012) A draw-and-guess game to segment images. In: SocialComm 2012 ASE/IEEE international conference on social computing, pp 914–917

9. Genevieve P (2012) SUN attribute database: discovering, annotating, and recognizing scene attributes. In Proceedings of the 2012 IEEE conference on computer vision and pattern recognition (CVPR), pp 2751–2758
10. Glaser BG, Strauss A (1967) Discovery of grounded theory. In: Strategies for qualitative research. Sociology Press, Mill Valley
11. Hossfeld T, Keimel C, Hirth M, Gardlo B, Habigt J, Diepold K, Tran-Gia P (2014) Best practices for QoE crowdtesting: QoE assessment with crowdsourcing. IEEE Trans Multimedia 16:541–558
12. Howe J (2006) The Rise of crowdsourcing wired. Wired Mag 14(06):1–6
13. Kennedy L, Hauptmann A (2006) LSCOM Lexicon definitions and annotations (Version 1.0). Computer science department, Paper 949. http://www.repository.cmu.edu/compsci/949
14. Liu D, Hua X-S, Yang L, Wang M, Zhang H-J (2009) Tag ranking. In: WWW 2009 Proceedings of the 18th international conference on world wide web, pp. 351–360
15. Mason W, Suri S (2012) Conducting behavioral research on Amazon's Mechanical Turk. Behav Res Methods 44(1):1–23
16. Morris RR, Dontcheva M, Gerber EM (2012) Priming for better performance in microtask crowdsourcing environments. IEEE Internet Comput 16(5):13–19
17. Murphy GL (2002) The big book of concepts. The MIT Press, Cambridge
18. Nie L, Yan S, Wang M, Hong R, Chua T-S (2012) Harvesting visual concepts for image search with complex queries. In: Proceedings of the 20th ACM international conference on multimedia (MM'12). ACM, New York, pp 59–68
19. Nielsen J (1994) Enhancing the explanatory power of usability heuristics. In: Proceedings of the ACM CHI'94 conference, pp 152–158
20. Nielsen J, Landauer TKA (1993) Mathematical model of the finding of usability problems. In: Proceedings of ACM INTERCHI'93 conference, pp 206–213
21. Nowak S, Rueger S (2010) How reliable are annotations via crowdsourcing: a study about inter-annotator agreement for multi-label image annotation. In: MIR'10 Proceedings of the international conference on multimedia, information retrieval
22. Paolacci G, Chandler J, Ipeirotis PG (2010) Running experiments on amazon mechanical turk. Judgment Decision Making 5:5
23. Parikh D, Grauman K (2011) Interactively building a discriminative vocabulary of nameable attributes. In: Computer vision and pattern recognition (CVPR), pp 1681–1688
24. Prelec D (2004) A bayesian truth serum for subjective data. Science 306(5695):462–6
25. Quinn AJ, Bederson BB (2011) Human computation: a survey and taxonomy of a growing field. Proc ACM CHI 2011:1403–12
26. Rafferty P, Hidderly R (2005) Indexing multimedia and creative works. Ashgate, Farnham
27. Ross J, Lilly Irani M, Silberman S, Zaldivar A, Tomlinson B (2010) Who are the crowdworkers?: shifting demographics in mechanical turk. In: CHI'10 Extended abstracts on human factors in computing systems (CHI EA'10). ACM, New York, pp 2863–2872
28. Smeaton AF, Over P, Kraaij W (2006) Evaluation campaigns and TRECVid. In: Proceedings of the 8th ACM international workshop on multimedia information retrieval (MIR'06) pp. 321–330
29. Snoek CGM, Freiburg B, Oomen J, Ordelman R (2010) Crowdsourcing rock n' roll multimedia retrieval. In: MM'10 Proceedings of the international conference on multimedia, pp. 1535–1538
30. Szabó ZG (2013) Compositionality. In: Zalta EN (ed) The stanford encyclopedia of philosophy. (Fall, Fall River). http://www.plato.stanford.edu/archives/fall2013/entries/compositionality/
31. Van den Haak MJ, Jong MDT, Schellens PJ (2004) Employing think-aloud protocols and constructive interaction to test the usability of online library catalogues: a methodological comparison. Interact Comput 16(6):1153–70
32. Van der Geest T, Spyridakis JH (2000) Developing heuristics for web communication. Tech Commun 47(3):359–82
33. Vliegendhart R, Larson MA, Pouwelse JA (2012) Discovering user perceptions of semantic similarity in near-duplicate multimedia files. In: CrowdSearch 2012: First international workshop on crowdsourcing web search, pp 54–58

34. Wieneck L, Düring M, Sillaume G, Lallemand C, Croce V, Lazzaro M, Nucci F, Pasini C, Fraternali P, Tagliasacchi M, Melenhorst M, Novak J, Micheel I (2013) Building the social graph of the history of European integration. A pipeline for humanist-machine interaction in the digital humanities. In: Proceedings of the conference histoinformatics 2013, Kyoto
35. Xirong Li; Snoek CGM, Worring M, Smeulders AWM (2012) Harvesting social images for bi-concept search. IEEE Trans Multimedia 14(4):1091–1104

Index

A
Action recognition, 187
AdaBoost, 58, 63–67, 74
Affect, 186, 189, 194
Agglomerative clustering, 53, 63–67
Amazon Mechanical Turk, 253, 256, 259, 266
Annotation, 186, 189, 191, 192, 199
Attributes, 260, 266
Audio concepts, 192

B
Bag of visual words, 187
Bag-of-words (BoW), 29, 30
Basic-level concept, 246
Bayesian fusion schema, 161, 166, 167, 175, 176, 180, 181
Bayesian networks, 166, 168, 175, 187
Bayesian Truth Serum, 253
Benchmark, 185, 187, 188, 200
Biomedical retrieval, 211, 219, 224

C
Case-based retrieval, 210, 211, 220–224
Coding, 29–32, 34, 36, 38–40, 44, 45
Complex interpretation, 231–238, 243–246, 248, 249, 252–254, 258, 260, 261, 266, 267
Compositionality, 234
Conceptual feedback, 67, 71, 73, 74
Connotation, 248, 249
Consensus, 239, 244–246, 249, 252, 257
Content extraction, 112, 115
Context, 229, 231, 239–243, 249, 253–255, 261, 266, 267

CrowdFlower, 250
Crowdsourcing, 229–231, 233, 240, 249–254, 256, 259–261, 265–267

D
Dataset, 185, 186, 188, 190
Decision tree, 135, 145, 153
Difference coding, 110, 113, 118, 122, 129

E
Early fusion, 200, 203, 204
Evaluation, 186, 187, 190, 193, 194

F
Feature extraction, 135–137, 142, 144–146, 151, 153, 156
Feature fusion, 3
Forensic system, 162–164, 176
Fuzzy classification, 174

G
Grounded theory, 260

H
Hierarchical fusion, 71
High-dimensional data, 136, 137, 142, 147, 150, 152, 155, 156
Human interpretations, 229–231, 235, 238, 241, 243, 244, 249–251, 253, 261, 265, 267

I

Image classification, 29–31, 35, 49
ImageCLEF, 215, 220, 222, 224
ImageCLEF photo annotation, 1, 8, 12, 25
ImageNet, 251
Information fusion, 210, 217

L

Late fusion, 53, 57, 58, 61, 62, 67, 70, 72, 74, 113, 123, 126, 129, 195, 197, 201, 203, 204
LSCOM, 237, 238

M

MediaEval, 185, 186, 188, 189, 193, 195, 205, 206
Microworkers, 250
Mid-level representation, 29–31, 33, 34, 36
Motion classification, 165, 166, 173
Motion features, 177
Movies, 185–190, 192, 203
Multimedia event detection, 110–112, 114, 117, 120, 125, 126, 129
Multimedia systems, 229–232, 243, 245, 249, 258, 267
Multimodal, 186, 189–191, 194, 196, 197, 201, 203, 204
Multimodal fusion technique, 161, 163–167, 170, 175–178, 180, 181
Multimodal surveillance system, 164, 170
Multimodality, 1–3, 18, 21, 23

N

Neural network, 198, 199, 205

O

Object classification, 161, 163–166, 170, 171, 176, 181
Object detection, 165, 172
Object recognition, 79–82, 84–87, 91, 94, 96–102, 104
Objectivity/objective, 230, 244, 245, 266

P

Pooling fusion, 31, 38
Presupposition, 242, 243

Q

Quality of Experience, 250

R

Rotation-based ensemble, 135–137, 143, 145, 147
Rule of Thirds, 249

S

Saliency, 79–81, 84, 86–92, 94, 95, 97–101, 103, 104
Security, 161, 162
Semantic concepts, 57, 72
Semantic indexing, 54, 56, 58, 68, 69
Shot segmentation, 193
SIFT, 197
Simple concept, 231–238, 240
Simple concepts, 244–246, 249, 252, 254
Social multimedia, 229, 230, 232, 240, 252, 253, 265
Subjectivity/subjective, 230, 238, 243–246, 248, 249, 266
Supervised classification, 195, 197
Support vector machine, 135, 137
Surveillance networks, 162–164, 170, 179, 180
SVM, 141

T

Tags, 231, 233, 237, 262–264
Temporal re-scoring, 67, 71–74
Textual feature, 1, 5, 8, 13, 15, 16, 19, 21, 25
TRECVid, 237

V

Video, 54, 56–58, 65, 67–70, 74
Video surveillance, 161, 162, 167, 177
Violence, 186–191, 193, 199, 201, 203
Violent scenes detection, 194
Visual attention, 79, 87, 89
Visual classification, 172, 173, 178, 180
Visual concept detection, 232
Visual concept recognition, 2, 3, 12, 25
Visual concepts, 188, 192
Visual feature, 2, 3, 5, 8, 9, 13, 14, 19–21, 163, 173, 178

W

Wearable video, 80, 91
World Health Organization, 189

Printed by Printforce, the Netherlands